The Biology of Hover Wasps

Stefano Turillazzi

The Biology of Hover Wasps

Springer

Stefano Turillazzi
Dep of Evolutionary Biology "L. Pardi"
Università degli Studi di Firenze
Firenze
Italy

Additional material to this book can be downloaded from http://extras.springer.com/

ISBN 978-3-642-32679-0 ISBN 978-3-642-32680-6 (eBook)
DOI 10.1007/978-3-642-32680-6
Springer Heidelberg New York Dordrecht London

Library of Congress Control Number: 2012955939

Printed on acid-free paper

Springer is part of Springer Science+Business Media (www.springer.com)

To Cristina, Valentina and Francesco

Foreword

The importance of this book is evident from a brief history of hover wasp science. When William Morton Wheeler wrote his landmark books on the social insects (1923, 1928), he devoted only two pages to the behaviour and natural history of the hover wasps (Stenogastrinae), even though he considered them (1923, p. 72) "a group of great interest, because they form a transition from the solitary to the social wasps". Wheeler's account of the hover wasps was based entirely on the work of F. X. Williams (1919) on just four species found in the Philippines. Williams described the remarkably diverse nests of these delicate forest insects, illustrating them with beautiful drawings that were tantalising invitations to further study because they suggested behavioural sophistication and social diversity. But they continued to be virtually neglected for decades. When the successor to Wheeler's books, E. O. Wilson's *The Insect Societies*, was published more than 50 years later (1971) even less space went to the hover wasps, and only two references (Iwata, 1967 and Yoshikawa et al. 1969) were added. Meanwhile, most research on the group consisted of taxonomic descriptions and scattered observations of nests and natural history (for a thorough history, see Chap. 1 of the present book).

This situation was to change with a quantum leap when Stefano Turillazzi, along with his professor and mentor Leo Pardi, went to Java in 1979 to seek the Indonesian stenogastrine wasps that had been described by the Dutch hymenopterist Jacobus van der Vecht some years before. Their focus on the hover wasps was the accident of a frustrated plan to go to Madagascar to study *Belonogaster*, another neglected taxon of social wasps. But the root of their interest in the two groups was the same: to explore the biology of little known wasps at the threshold of sociality.

The present book carries forward the spirit of the work that began, in 1979, with the combined talents of Pardi and Turillazzi. Both were already noted for their skill and patience as insightful observers of behaviour and natural history in the field. And both melded an interest in theoretical issues with a desire to examine evolutionary questions using data on natural populations, supplemented by experimental tests and dissections. They soon discovered that not only did some species have a clear dominance hierarchy and a reproductive division of labour between an alpha female (the exclusive egg layer) and subordinate helpers, but they also observed the construction and functions of ant guards and of the unique gelatinous material placed under eggs and near small larvae, all subjects of speculation in the early literature.

During the decades that have followed the Pardi-Turillazzi expedition to Java, Stefano Turillazzi and his students and colleagues have gone on to deepen studies of several other social wasps, especially of *Polistes* and its social parasites. They have formed that rare and productive thing, a *school* of research and theory on the social wasps that has made their university in Florence a world famous centre for research on social insects. I have heard it said of Stefano Turillazzi that while tourists may go to Florence to see Michelangelo's *David,* students of social insects go to Florence to see Turillazzi and his associates, a group that still actively pursues a fine tradition of research on the social wasps. Many of that group are named in the Preface to this book. Although the laboratories devoted to social insect research will soon move to a new building on the outskirts of Florence, not the least of the charms of their long-time location in the centre of the city next to the Palazzo Pitti was the chance to see caged hover wasps and colonies of *Polistes* thriving in rooms overlooking the ancient rooftops of the city, high in the tower that houses the abandoned eighteenth century observatory that gives the museum of natural history, *La Specola*, its name.

The breadth of interests of the Florence group and of Stefano Turillazzi himself is reflected in this book. It contains an ambitious survey of ideas about the evolution of the social insects and an assessment of the current standing of controversial concepts. But the most remarkable and special characteristic of the book, and the aspect that will best stand the tests of time, is that is represents the culmination of a lifetime of work on a single fascinating group of insects. It belongs to a rare and uniquely valuable *genre* of scientific literature: a synthesis that can only come from a passion for a particular group of organisms that embraces every aspect of their biology and behaviour and is based on first-hand experience in the field. For that enviable contribution of this book, students of social insects will be, literally, eternally grateful.

July 2012 Mary Jane West-Eberhard

Preface

Here, in a small bungalow on the southern border of the Genting Tea Estate, on the last day of my 3-week study period and with my eyes still full of the colours and shapes of a sunny day spent in the surrounding forest, I thought it was the perfect moment to write a short presentation to this book. Genting Tea Estate (or GTE according to a Malaysian custom to indicate places with their initials) is a lovely area of secondary growth forest, property of an English amateur naturalist, Mr Henry Barlow, well known in science as an authority on East Asian moths. The estate is located on the pass crossed by an ancient road connecting the West coast to the tin mines and by the modern Leburahia Pantai Timur (the East Coast Highway). Here in this ground, I have spent several weeks almost every year since 1984, studying hover wasps, Stenogastrinae for science and "pegnangat" for the village people.

Now that I am not sure whether I shall be able to come back again next year because of the critical situation of research funding in Italy, it is even more obvious to me how important this book is in providing a summary of the knowledge acquired on these wasps over the past 32 years. Looking at the hills on the other side of the valley, that year after year have become barer and barer as the forest has slowly been replaced by plantations of green beans, subjected to massive pesticide treatments, I wonder how long this area will be suitable for hosting insect populations. In particular for these wasps, one of the infinite essences of a forest which never ceases to furnish surprise after surprise to anyone willing to observe and study it.

Yes, I think this book is necessary—it will leave future generations an outline of the biology of a group of insects which still represent, from many aspects, an enigma of the evolution of social life. In its seven chapters, the book traces a trail that, starting from a concise presentation of these wasps and of the researchers who mainly contributed to their study, weaves through the description of the principal characteristics of morphology, anatomy of the adults and development of larvae: a subject still full of questions to which it is difficult to reply. The following chapters review the main aspects of the behaviour of males and females, in comparison with that of other species of social wasps. The characteristics of the social organisation of the very few species in which this has been studied are then considered and commented. Communication is the main characteristic by which social relationships between individuals are possible in a wide range of organisms. This aspect is treated in a special chapter in which the different information channels found in these insects are reviewed. But these insects differ from all the other members of the social club for the architecture of their nests, a special character that makes it easier to recognise different species in the field. This is treated in the longest chapter of the book with various references to the better known species. Finally, the last chapter, after a concise exposition of the fundamental problems regarding the evolution of social life in insects, tries to summarise the main steps in the evolution of sociality in these wasps, furnishing a personal interpretation that sometimes matches and sometimes contrasts those of friends and colleagues.

The information reported in the volume is the result of research which began over a century ago and of work by scholars and students from various countries. I think that the contribution from my group has been important, as well as that of English, American and Japanese researchers, but recently Malaysian and Vietnamese entomologists have also begun to show an interest in the study of these insects. This is very important. Indeed, it will be the younger generations of those countries where these insects live who will find the answers to a whole range of yet unsolved problems regarding the biology of these wasps and contribute to the systematic description of new species.

Writing this book would never have been possible without the help of many students of mine who, mission after mission, gave more or less important and more or less successful (in terms of publications) contributions to the research. Some of them have continued to work in the scientific field, but for most spending some time in the context of scientific work in our team has remained a sort of life experience that, I hope, has brought them other interior rewards. I am particularly indebted to Rita Cervo, Matthew Sledge, Elisabetta Francescato, Daniele Fanelli, Christina Coster-Longman, Monica Landi, Angelo Fortunato, Francesca Romana Dani, Leonardo Dapporto, Duccio Lambardi, Irene Ortolani, Iacopo Petrocelli and, more recently, to David Baracchi who is the student who has spent most time in Malaysia, producing important contributions to the study of several species. Other students and researchers have helped me in Italy by studying the material collected during the various missions. I have to thank in particular Gloriano Moneti and Giuseppe Pieraccini of the Mass Spectrometry Centre of the University of Florence (C.I.S.M.) who have been reference points for me and my students for studies performed with Mass Spectrometry techniques.

A particular acknowledgment goes to my Malaysian friends and colleagues who in the course of the years have become more and more interested in collaborations, giving important contributions to research, including logistic and organisational aspects. First of all professor Rosli Bin Hashim, at present director of the Institute Sains Biologi of the Universiti Malaya, and also Prof. Yong Hoi Sen and Prof. Sofian Azirun of the same University. Special thanks go to Mr. Henry Barlow who, beyond his logistic support and personal encouragement in writing this book, revised and corrected the very first version of the piece of writing destined to become the basis of the book. Mr. Hok Kim Loong has been of invaluable help in the field and often solved various kinds of problems for us.

I must also thank my colleagues and friends who reviewed corrected and discussed the different chapters, offering important suggestions. First of all my friend (but also scientific guide) Mary Jane West-Eberhard who discussed the general plan of the work with me as well as the critical first and last chapters of the book and accepted to write the Foreword. Mike Hansell, who was my fellow in Papua New Guinea searching for some of the most elusive hover wasps, as one of the world authorities on animal construction reviewed the longest chapter of the book, that on nest architecture.

Ragavendra Gadagkar reviewed and commented on the last chapter and that on behaviour that was also reviewed by Joan Strassmann.

Jeremy Field, who with his group has provided important findings on the sociobiology of *Liostenogaster flavolineata* over the last few years, reviewed and commented the chapter on Colonial Dynamics.

An important contribution to the revision of the chapters on the morphology and anatomy and on the systematic position of the Stenogastrinae came from James Carpenter, reference authority on the systematics and phylogeny of the Vespidae, who also read and commented the concluding chapter furnishing stimulating counter opinions about the evolutionary route to sociality of these wasps.

Francesca Romana Dani reviewed the chapter on Communication.

Chris Starr, finally, also responsible for the Archive of the International Union for the Study of Social Insects (IUSSI), reviewed and commented the paragraphs on the story of the research on hover wasps.

Particular thanks are due also to my ex-student Christina Coster-Longman, who was with me in various missions to Malaysia and who reviewed the final version of the English text, and to David Baracchi who helped me in editing the bibliography and tables.

The photographs, pictures and figures are the result of the work of various students and technicians who were my fellows during various study campaigns in Java, Peninsular Malaysia and Sabah and Papua New Guinea. Mr. Riccardo Innocenti, photographer of the Zoological Institute of the University of Florence, was the very first, followed by Mr. Saulo Bambi. Credits for the various pictures are reported, when possible, in the respective legends.

Funds for various research campaigns were furnished by Università degli Studi di Firenze, the Italian Council of Research (C.N.R.), Italian Ministry of Education and Research (M.U.R.S.T.) and TMR Network "Social Evolution" project of the European Commission. The Italian Foreign Ministry financed two fellowships for Malaysian students who carried out joint researches on these wasps after or during a training period spent in our laboratories.

Finally, last but not least, I want to thank my family for their patience and support and for the understanding of my work and interests during the past years.

I would like this book to be only a starting point for future studies on these insects that, I hope, will be carried on by young scientists of the countries where hover wasps live and are an important component of the entomological fauna. But the hover wasps are not only one of the infinite groups of insects populating South East Asian forests but also are somewhat special for the characteristics of living together and represent, perhaps, a unique output of social evolution.

11 Feb 2012 Stefano Turillazzi

Contents

Chapter 1
The Hover Wasps

1.1 The Stenogastrinae: Genera and Species

Stenogastrine wasps, the subject of this book, are a group of wasps common in some South East Asian countries. They have been included as a subfamily in the family Vespidae (Carpenter 1982). Their morphology and biology, however, present striking differences with the other subfamilies of social wasps (Polistinae and Vespinae) and recently their phylogenetic position has been the argument of strong debate (Schmitz and Moritz 1998; Carpenter 2003; Hines et al. 2007; Pickett and Carpenter 2010) (Fig. 1.1). According to the very latest cladistic elaboration of morphological, behavioural and biomolecular characters by Pickett and Carpenter (2010) (which confirms that of Carpenter 1982; 1988), the hover wasps should be considered as the sister group of the social vespid subfamilies (Polistinae + Vespinae) (Fig 1.1a).

Concerning the taxa which form the subfamily and their respective relationships, the most recent cladogram, proposed by Carpenter and Starr (2000), recognises seven genera distributed in the Oriental and Papuan Regions and is reported in Fig. 1.2.

Eustenogaster is the most widely distributed, being reported from South India to Indochina, Indonesia and the Philippines. *Cochlischnogaster* has been found from South China to Indochina, *Metischnogaster* is reported from Sumatra, Borneo, Peninsular Malaysia and the Philippines. *Liostenogaster* and *Parischnogaster* range from South China to great part of Indonesia and Philippines, but some species of the second genus have also been found in Assam and Sikkim. *Anischnogaster* and *Stenogaster* live only in New Guinea and nearby islands (Fig. 1.3).

At present, only *Anischnogaster*, *Stenogaster*, *Metischnogaster* (van der Vecht 1972, 1975, 1977) and *Eustenogaster* (Saito and Kojima 2007) have been revised.

S. Turillazzi, *The Biology of Hover Wasps*, DOI 10.1007/978-3-642-32680-6_1,

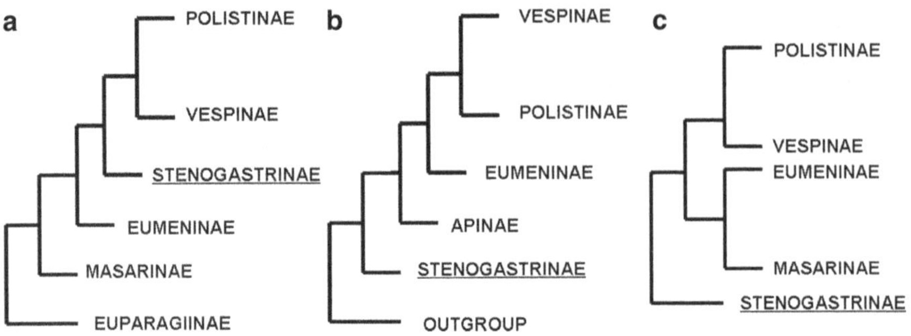

Fig. 1.1 Three proposed cladograms showing the possible phylogenetic relationships of Stenogastrinae. (**a**) Carpenter (1982, 1988); Pickett and Carpenter (2010). (**b**) Schmitz and Moritz (1998). (**c**) Hines et al. (2007)

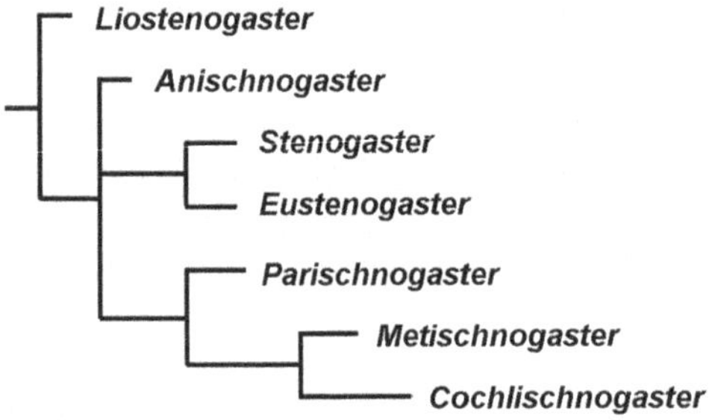

Fig. 1.2 Phylogenetic relationships of the genera of the subfamily Stenogastrinae (from Carpenter and Starr 2000, modified)

Parischnogaster is probably the largest and least known genus, where about two-third of the species remain undescribed (according to a personal communication from Chris Starr).

A total of 58 described species are known at present, mainly pertaining to the genera *Liostenogaster*, *Parischnogaster*, *Stenogaster* and *Eustenogaster*. The names and authors of the various species in the seven genera (from Carpenter and Kojima 2006) follow. References to the original papers can be found in the same paper.

Anischnogaster van der Vecht 1972 (5 species): ***A. dubia*** van der Vecht 1972, ***A. iridipennis*** (Smith) 1859, ***A. laticeps*** van der Vecht 1972, ***A. loriai*** *(A. l. loriai* (du Buysson) 1909, *A. l. maculata* van der Vecht 1972), ***A. spilaspis*** (Cameron) 1913.

Cochlischnogaster Dong and Otsuka. 1997 (3 species): ***C. daduganensis*** Dong and Otsuka 1997, ***C. menglunensis*** Dong and Otsuka 1997, ***C. spatulata*** (Carpenter and Starr) 2000.

Eustenogaster van der Vecht 1969 (15 species): ***E. agilis*** (Smith) 1860, ***E. calyptodoma*** (Sakagami and Yoshikawa) 1968, ***E. eximia*** *(E. e. eximia*

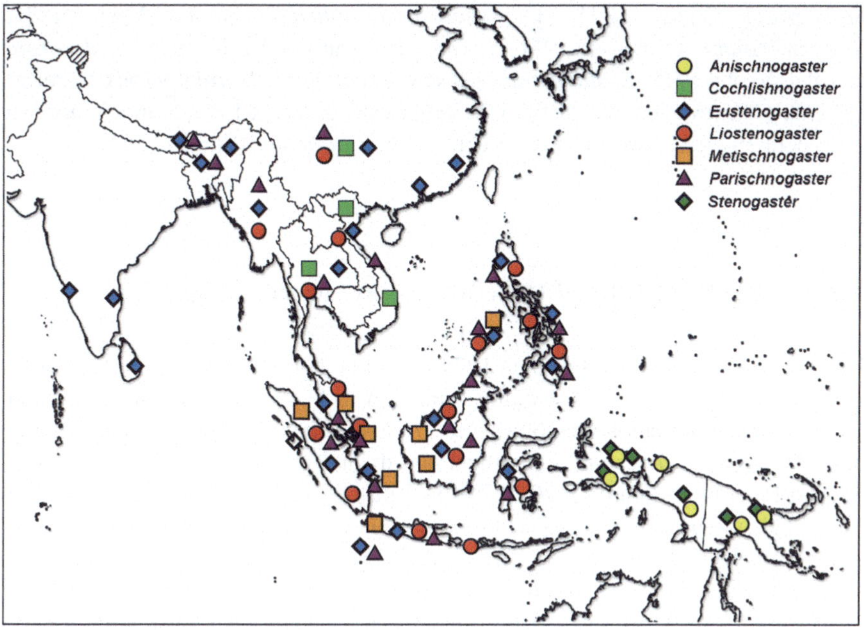

Fig. 1.3 Approximate geographical distribution of the Stenogastrinae genera. Localities are taken from Carpenter and Kojima (2006)

(Bingham) 1890, *E. e. eximioides* (Dover and Rao) 1922), ***E. fraterna*** (Bingham) 1897, ***E. fulvipennis*** (Cameron) 1902, ***E. fumipennis*** Saito 2007, ***E. gibbosa*** Starr and van der Vecht 2006, ***E. hauxwellii*** (Bingham) 1894, ***E. latebricola*** Saito 2007, ***E. luzonensis*** (Rohwer) 1919, ***E. micans*** (De Saussure) 1852, ***E. nigra*** Saito and Nguyen 2006, ***E. palavanica*** Reyes 1988, ***E. scitula*** (Bingham) 1897, ***E. spinicauda*** Saito 2007.

Liostenogaster van der Vecht 1969 (12 species): ***L. abstrusa*** Turillazzi 1999, ***L. campanulae*** Turillazzi 1999, ***L. filicis*** Turillazzi 1999, ***L. flaviplagiata*** (Cameron) 1902, ***L. flavolineata*** (Cameron) 1902, ***L. nitidipennis*** (De Saussure) 1853, ***L. pardii*** Turillazzi and Carfì 1996, ***L. picta*** (Smith) 1860, ***L. topographica*** Turillazzi 1999, ***L. tutua*** Turillazzi 1999, ***L. varipicta*** (Rohwer) 1919, ***L. vechti*** Turillazzi 1988.

Metischnogaster van der Vecht 1977 (2 species): ***M. cilipennis*** (Smith) 1857, ***M. drewseni*** (De Saussure) 1857.

Parischnogaster von Schulthess 1914 (10 species): ***P. alternata*** Sakagami 1969, ***P. aurifrons*** (Smith) 1862, ***P. depressigaster*** (Rowher) 1919, ***P. gracilipes*** (van der Vecht) 1977, ***P. jacobsoni*** (du Buysson) 1913, ***P. mellyi*** (De Saussure) 1852, ***P. nigricans*** (Cameron) 1902, ***P. striatula*** (du Buysson) 1905, ***P. timida*** (Williams) 1910, ***P. unicuspata*** Reyes 1988.

Stenogaster Guerin 1831 (11 species): ***S. adusta*** van der Vecht 1975, ***S. canaliculata*** (Cameron) 1911, ***S. concinna*** van der Vecht 1975, ***S. decorata*** van der Vecht 1975, ***S. flavifrons*** van der Vecht 1975, ***S. fulgipennis*** (Guérin) 1831, ***S. glabra*** van der Vecht 1975, ***S. licina*** van der Vecht 1975, ***S. macilenta*** van der Vecht 1975, ***S. pilosa*** van der Vecht 1975, ***S. unicolor*** (Smith) 1864.

1.2 A Brief History of the Studies on Hover Wasps

My story with these wasps began in 1979 when Leo Pardi, my mentor and scientific guide decided to organise a research campaign in Madagascar for studying other social wasps, those belonging to the genus *Belonogaster*. The observations of Pardi on *Belonogaster grisea* in Somaliland (Pardi and Marino-Piccioli 1970, 1981; Marino-Piccioli and Pardi 1970; 1978) had inspired a new interest in my master for the behaviour of social insects, after quite a long period dedicated to research on astronomical orientation of sand fleas (Talitridae). Madagascar, rich in endemic species, promised to be a good territory where we expected to clarify some aspects of the biology of those wasps. Unfortunately, refusal from the local authorities to issue research visas in a timely manner compelled us to change our plans so as not to lose already allocated funds. Pardi suggested, to my great satisfaction, to turn his interests towards the Orient, with a preliminary project on wasps which various authors had indicated as a quite promising field for the study and survey of the origins of social behaviour in insects: the Stenogastrinae. Pardi's interests had already focussed on the problems of the origin of insect sociality and just a few months before he had discussed the definitions of various social stages in a conference at the Academy of Lincei in Rome, giving a clear synthesis of the evolutionary steps which, as suggested by the comparative method applied to living species, seemed to characterise the progression from solitary forms to the highly social ones in the Hymenoptera (Pardi 1980). So turning our attention to the hover wasps, where, according to previous but scarce scientific reports, solitary and social characteristics seemed to intertwine, became obligatory. Choosing the country where to develop this research was problematic as the distribution area of these wasps covers the greater part of South East Asia and New Guinea, moreover very little information was at that time available on the actual distribution of species and colonies. Eventually it was decided to visit one of the better known scientific institutions in the area: the Botanical Garden of Bogor, on Java island, Indonesia; thus we prepared the mission for the end of the year, first searching for all the available information on these wasps. We found the first report on stenogastrine wasps in a picture in the Atlas by Guérin-Méneville (1831) who illustrated the first known species, *Stenogaster fulgipennis*, among the material collected during the expedition of the ship Coquille to New Guinea.

As van der Vecht recounts (van der Vecht 1975), Guérin later changed the name of the genus to *Ischnogaster* as he feared that *Stenogaster* had been previously used

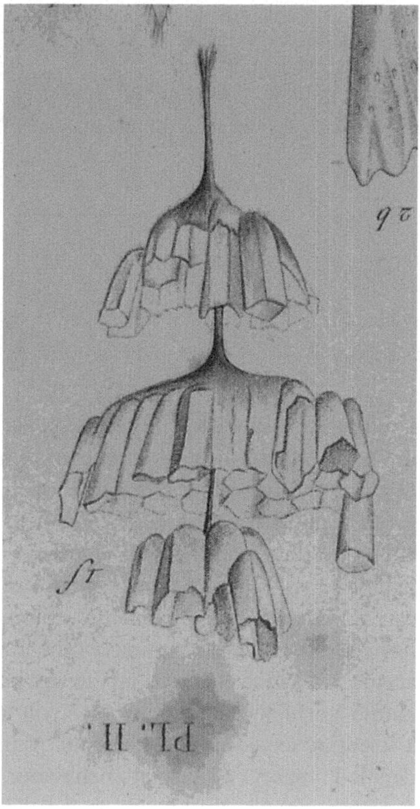

Fig. 1.4 Plate II of "Etudes sur la famille des Vespides" by H. De Saussure (1852–1858) with the drawing of the nest of *Parischnogaster mellyi* (upside down in the original plate)

before publication of the text of his paper in 1838, but the validity of the former name was eventually confirmed (Dover and Rao 1922).

The first author who treated their systematic position was H. De Saussure in his monumental work "Etudes sur la famille des Vespides" (1852–1858). In Plate II the author provides a drawing, upside down, of a nest of *Parischnogaster mellyi* (Fig. 1.4). Considering their place with respect to the subfamilies of Eumeninae and Vespinae he remarked that these wasps were, in all their characters, entirely intermediate between the two tribes. Hence they were ascribed to the social wasps only according to their habits. Williams (1919, 1928) in the Philippines, provided the first information on their behaviour and sketched the peculiar nest architecture of some species.

Until 1927, when A. von Schulthess created the new genus *Parischnogaster* for some species living in the Oriental Region, all species were placed under the single genus *Stenogaster*. Descriptions of other species ensued during the following decades especially using museum material, with the exception of a note by the Dutch entomologist Edward Jacobson (1935) on two species of *Parischnogaster* observed in Java, and of studies led in the field, in Malaysia, by H.T. Pagden (1958, 1962). The latter realised the wide variety of nest architecture of these wasps and its

Fig. 1.5 The author with Jacobous van der Vecht in front of his house in Putten, Holland, in 1984

importance in species differentiation and tried to furnish the first organic division of the group based on nest characteristics. Pagden was also the first to notice and describe the patrolling behaviour of males of two species of *Metischnogaster*, a genus established by van der Vecht in 1977. Also some Japanese entomologists begun to show interest in these insects; first of all Kunio Iwata who wrote a paper, published in 1967, on the peculiar habits of hover wasps, mainly based on the careful analysis of specimens, colonies and nests collected in Thailand. He furnished the first descriptions of the immature stages of various species, providing accurate drawings of eggs, larvae and pupae. A group led by Shôichi Sakagami and Kimio Yoshikawa and joined by Ryoh-ichi Ohgushi and Sôichi Yamane, undertook a scientific expedition in Thailand, Malaysia and Cambodia in 1966, with the first detailed report of social behaviour observed in a species of the genus *Parischnogaster* (Yoshikawa et al. 1969). We now come to the papers of Jacobous (Jack) van der Vecht, a Dutch entomologist who spent much of his life in south East Asia around the time of the Second World War, for revisions of the taxonomy of the group. Van der Vecht created five new genera including species from the entire distribution area and preliminarily described tens of new species. He described the material collected during the 1966 Japanese campaign and named the genera *Liostenogaster* and *Eustenogaster* in the report of the expedition (Yoshikawa et al. 1969). Then he revised the two Papuan genera *Anischnogaster* and *Stenogaster* (van der Vecht 1972, 1975) and the oriental genus *Metischnogaster* and created a seventh genus, *Holischnogaster* which was then synonymised by Carpenter (1982) with *Parischnogaster*. At the end of his scientific life, he left his notes to younger specialists to conclude the revision of the other three largest genera of the oriental region: *Liostenogaster*, *Eustenogaster* and *Parischnogaster*. I was one of those young entomologists when in 1981, and later in 1984, I visited him at his house, named symptomatically "*Andrena*", in Putten, Holland (Fig. 1.5).

In 1975 another wasp specialist entered the scene: John Philip Spradbery published an account on the biology and nest architecture of a Papuan hover wasp, *Stenogaster concinna*, in the genus freshly revised by van der Vecht (Spradbery 1975). But, unfortunately, further contributions of this authority in the studies of hover wasps resulted only in one more paper (on *Anischnogaster*),

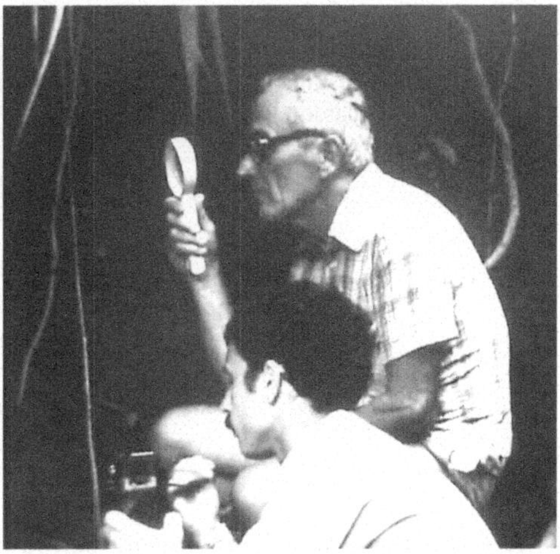

Fig. 1.6 Leo Pardi, assisted by the author, during the first expedition to Java in 1979 (photo by Riccardo Innocenti)

Fig. 1.7 Mike Hansell from Glasgow University

published almost 15 years later (Spradbery 1989). Also the basic social biology and the nest of a species of *Eustenogaster*, *E. eximia*, was the subject of a short note by the American entomologist Karl Krombein published in 1976 (Krombein 1976).

So, in September–October 1979, L. Pardi, his wife, me and an expert photographer, Mr Riccardo Innocenti, arrived in Bogor and settled down in Tegal Lega, a small village in the neighbouring town (now encompassed in the metropolitan area of the city). I still remember the very first hover wasp we discovered in an earth trench near the village: a small initial nest of *Parischnogaster* with only one female (Fig. 1.6).

Just before leaving Italy we learned that a British zoologist from Glasgow University, Michael Hansell (Fig. 1.7), had recently finished a preliminary survey in Thailand on *Parischnogaster mellyi* and we suddenly realised that the most common species in our area was the very same. However we found another species in the surroundings, very similar in appearance but with a quite different nest (that was later determined by van der Vecht as *Parischnogaster nigricans serrei*), and we decided to focus on that to avoid possible overlapping with Hansell's research.

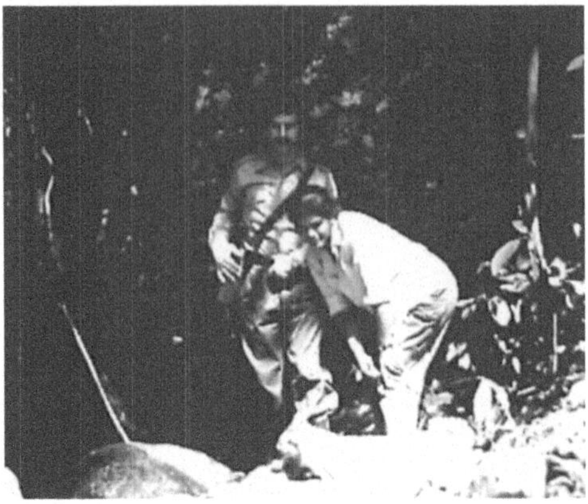

Fig. 1.8 Charlotte Samuel with the author during a trip in the Malaysian forest in 1984 (photo by Riccardo Innocenti)

It was actually a very interesting period spent with Pardi in the small village, observing colonies found on threads under the roofs of that "Indonesian *favela*" surrounded by dozens of children who simultaneously became our "disturbers" and "helpers". I visited the same place two years later, together with my wife Cristina, when we spent three and half months focusing on the study of the colonial cycle of the same species.

So these years signified a true and proper start up for research on hover wasps: Mike Hansell in Thailand, me in Java while the Japanese, especially R. Ohgushi, re-activated their research projects on Stenogastrinae with new missions to Sumatra, starting from 1982.

In 1981 Mike established contacts with the University of Malaya in Kuala Lumpur and began new research with the Malaysians on a species of *Liostenogaster*, *L. flavolineata*. Its social biology and natural history was the subject of the PhD thesis of a Malaysian student, Charlotte Samuel, who performed a really great job studying this species for several years in the surroundings of Kuala Lumpur and producing the most complete survey, at present, on the social biology of a stenogastrine wasp (Samuel 1987) (Fig. 1.8).

I met Mike and Charlotte at the Meeting of the International Union for the Study of Social Insects (IUSSI) in August 1982 in Boulder and I was really interested in research possibilities in Malaysia; so the year after, endorsed by Mike himself, I contacted Prof. D.R.Wells and Prof. Yong Hoi Sen at the University of Malaya. At the end of 1984, for the first time I visited the Gombak Field Centre, 20 km North East of Kuala Lumpur, where I began to observe other species. Contacts and collaboration with Malaysian researchers became stronger over the following years, especially those with Prof. Rosli Bin Hashim, leading to the publication of several joint papers (Fig. 1.9).

In Malaysia I found the support of an English resident, Henry Barlow, quite famous for his interests in Natural History and a leading expert on South Asian moths, who several times hosted me and my students in his property on the road to Genting Highlands.

Fig. 1.9 The author with two researchers of the University of Malaya, Sofian Azirun (*on the left*) and Rosli Bin Hashim (*on the right*)

Fig. 1.10 James Carpenter from American Museum of Natural History

At almost the same time an American entomologist, Chris Starr, after his appointment as associate professor at a Philippine university in 1981, performed collections and observations on various species of hover wasps. Unfortunately Chris mainly published only brief, if interesting reports, on his research (Starr 1984) and he is still responsible for a revision of the genus *Parischnogaster* which he has left, for the moment, unfinished.

But 1981 is also an important year for the first publication which took into account the phylogenetic revision of the superfamily Vespoidea; James Carpenter of Cornell University (Fig. 1.10) treated the division and evolutionary relationships of the subfamilies of the Vespidae in particular and regarded the Stenogastrinae as the sister group of other social wasps, Polistinae + Vespinae, contrary to opinions expressed by some previous authors. Richards (1971), van der Vecht (1977) and Spradbery (1975) followed by Pardi and myself (Pardi and Turillazzi 1982) had argued, in fact, that the hover wasps presented so many morphological and behavioural differences with Polistinae and Vespinae that a group composed of these subfamilies could not be considered as monophyletic and stressed the similarity between Eumenidae (sensu Richards) and Stenogastrinae. In particular, van der Vecht (1977) counted eight main morphological differences between Stenogastrinae and Polistinae + Vespinae (for example, unlike the other Diploptera, the hover wasps do not fold their wings along the length of their back, neither do females present glands on the two last gastral sternites as in Polistinae and Vespinae) and pointed to the similarities these wasps share with the Zethinae, a group of the Eumenidae, concluding that the Stenogastrinae should be considered as a subfamily of the Eumenidae or as a distinct family. In this respect two Indian entomologists, B.P. Das and V.K. Gupta, had published in 1984 "a catalogue of the families Stenogastridae and Vespidae from the Indian subregion" following this opinion (Das and Gupta 1984). But if, according to Carpenter (1982), we use cladistic methods it appears that these characters are autoapomorphies of the Stenogastrinae and thus uninformative about the phylogeny of the group, while consistent synapomorphies shared by the three taxa confirm that Stenogastrinae, Polistinae and Vespinae are sister groups and derive from a common ancestor. The same author proposed organisation of the family Vespidae into six subfamilies: Euparaginae, Masarinae, Eumeninae, Stenogastrinae, Polistinae and Vespinae.

In the 1980s, various reports of research undertaken on different species belonging to three genera, *Parischnogaster*, *Liostenogaster* and *Eustenogaster* provided our knowledge of the basic biology of hover wasps, especially on larval development and rearing, colony size and nesting preferences, as well as behavioural repertoires; but no new species were described, with the important exception of one new *Eustenogaster* and one new *Parischnogaster* published in a review of Philippine Stenogastrinae by Stephen Reyes in 1988. In particular, we observed that all the species studied presented the same characteristics of very simple eusocial organisation. This was also true for a species of *Metischnogaster*, the fourth Oriental genus, which I briefly studied from 1985 to 1989. This was one of the reasons why Mike Hansell and I decided to also have a look at the other two genera about which information was very scarce: *Stenogaster* and *Anischnogaster*. Ethological information was limited to the observations on *S. concinna* by Spradbery (1975) who stated in his paper that this species "*displays the most primitive subsocial condition in which a single female, without cooperation or assistance from its progeny, cares for immature stages*". In October–December 1989 Mike and I visited the North coast of Papua New Guinea where we found a number of colonies of species of *Anischnogaster* and *Stenogaster* (Fig. 1.11). The results of our survey were in line with the opinion that basic eusocial elements were present in all the genera, constituting a unifying character for all hover wasps and already occurring in a common ancestor of the group.

In 1990, the Japanese group of entomologists led by S. Sakagami (Fig. 1.12) and R. Ohgushi resumed their research on Sumatran hover wasps in various chapters of a book published by the Hokkaido University Press; in particular the ethogram of *P. mellyi* and nest architecture of various species were reported in detail. At almost the same time I began the systematic analysis of species belonging to the genus *Liostenogaster*; using part of the important information found in van der Vecht's

Fig. 1.11 Mike Hansell and the author in Papua New Guinea in 1989 (photo by Saulo Bambi)

Fig. 1.12 Shôichi Sakagami

notes. I first described as new species *L. vechti* (Turillazzi 1988) that I had studied in Malaysia.

In August 1994 I attended another meeting of the IUSSI in Paris and there I met a young British entomologist who asked me many questions about the species and nidification sites of hover wasps before initiating new research on their social behaviour. Jeremy Field (Fig. 1.13) and his group began their work in 1995 on colony aggregations of *L. flavolineata*, in the same places where Charlotte Samuel worked, using an approach never before applied to these wasps. Their research was

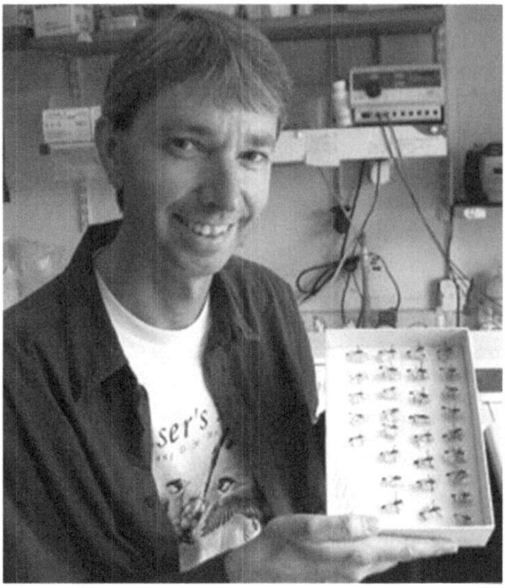

Fig. 1.13 Jeremy Field from the University of Sussex

based on precise questions, experimental manipulation of colonies and individuals and the utilisation of biomolecular techniques to ascertain relatedness parameters and to determine reproductive skew within groups of adults and larvae. This new line of research, which still continues with stimulating results, clearly demonstrated that hover wasps can be excellent models for studying the factors which led to the evolution of insect sociality.

The work of many students of mine and the J. Field research groups cannot be forgotten: from 1994 to present times a number of young scientists made important contributions. Among the most important ones I want especially to acknowledge Elisabetta Francescato, Matthew Sledge, Christina Coster-Longman, Angelo Fortunato, Monica Landi, Daniele Fanelli, David Baracchi, Leonardo Dapporto, Rita Cervo, Francesca Romana Dani, Duccio Lambardi, Irene Ortolani and my wife Cristina Marucelli; for Field's group, Seirian Sumner, Gavin Shreeves, Maurizio Casiraghi, Alan Bolton, Catherin Bridge, Michael Cant, and Adam Cronin.

In 1994, collaboration with the group of Joan Strassmann and David Queller of the Rice University had already led to the estimate of inbreeding and relatedness in colony members in *L. flavolineata* and *P. alternata* using allozyme variation (Strassmann et al. 1994). This collaboration continued in the following years with contributions to the study of a species of the genus *Eustenogaster* (Landi et al. 2003).

From 1995 to the present, studies on the biology and social behaviour of hover wasps were almost completely shared by the Italian and British groups with over 45 publications. Studies on physiological and anatomical aspects were, on the contrary, quite scarce with the exception of a new line of research begun by my group on chemical communication, directly connected with behavioural studies.

Concerning contributions to the study of systematics and phylogeny of the group, Jim Carpenter in 1988 published an important paper focusing on hover wasps which reported the relationships between genera inside the subfamily. In this paper Carpenter confirmed Stenogastrinae as the sister group of the other social wasps notwithstanding the fact that some authors (referring especially to me and Pardi) still continued to consider these wasps as something different from other social wasps, and stressing a single origin of eusociality for all the vespids. These opinions have twice been challenged again up to the present. The German authors Schmitz and Moritz (1998) disputed Carpenter's position and after cladistic analysis of biomolecular characters concluded that the Stenogastrinae cannot be placed near the Vespids but exhibit strong differences with them. This conclusion was rebutted by Carpenter and Starr (2000) and Carpenter (2003). More recently, a group of North American researchers (Hines et al. 2007) again challenged Carpenter's ideas in a paper which reports phylogenetic analyses of the family Vespidae based on data from four nuclear gene fragments. According to this analysis, the Stenogastrinae appear to be a clade distantly related to Polistinae + Vespinae, hence two independent origins of eusociality in Vespidae are indicated. The same concept was supported by Hunt (2007). But the story is not yet finished: Kurt Pickett and Jim Carpenter, in a recently published paper, counter point after point the arguments of Hines and co-workers confirming the phylogenetic position of hover wasps according to Carpenter's tree (Pickett and Carpenter 2010) and the monophyly of wasp eusociality. I shall come back to this story again at the end of the book, in Chap. 7.

In 1996, Carpenter and Kojima furnished the very first checklist of the species in the subfamily Stenogastrinae, after von Schulthess (1927) and Iwata (1971), giving 45 species in seven genera. In 2000 the subfamily had been augmented by another genus, *Chalogaster*, which Carpenter and Starr (2000) described based on specimens collected during an expedition to Vietnam by the American Museum of Natural History in 1998, while patrolling along a stream near the famous Ho Chi Min Trail. However, Carpenter (2001) had to synonymise *Chalogaster* with *Cochlischnogaster*, a genus described for two species found in Yunnan Province and published in Chinese just 3 years before (Dong and Otsuka 1997). The genus now includes three species, and seems to be the sister-group of the genus *Metischnogaster*, and is characterised by males with peculiar antennae (see Fig. 2.5).

Continuing the description of species belonging to the genus *Liostenogaster* I described six more species from Malaya (Turillazzi and Carfi 1996; Turillazzi 1999), while the genus *Eustenogaster* has recently been revised by Japanese entomologists (Saito and Kojima 2007). Junichi Kojima also published a study of hover wasp larvae (Kojima 1990) and more recently Japanese and Vietnamese researchers began interesting studies on species inhabiting the Northern part of the distribution area that have developed a tendency to hibernate (Saito et al. 2006, 2009). Phenetic studies on hover wasps were also performed by my group with the main contribution by David Baracchi, who began to investigate venom peptides and morphological similarities of species of various genera using innovative techniques (Baracchi et al. 2010, 2011).

1.3 The Other Social Wasps

In its widest sense, "wasp" is a name given to various Hymenoptera belonging to different families. They present a wide range of social adaptations but eusociality is concentrated in the Vespidae family. The Vespidae, in turn, comprise subfamilies with purely solitary species such as the Euparaginae and others with exclusively eusocial species, like the Polistinae and Vespinae. The Eumeninae show an interesting evolution toward sociality but, at present, no truly eusocial species has been found. Until a few years ago, the Stenogastrinae (Fig. 1.14) were thought to include both solitary and eusocial species; as we shall see in the course of this book, recent studies have shown that sociality has assumed particular aspects in the subfamily. Studies on the social evolution of Vespids are numerous (to mention only a few: Evans and West-Eberhard 1970; Jeanne 1980; Pardi 1980; Carpenter 1991; Ito 1993; Gadagkar 2001; Hunt 2007, 2011), and, as Wilson (1971) observed, these insects are those in which three of the main characteristics of social insects have been discovered: trophallaxis (the transfer of food from one member to another in a colony), social dominance and nutritional control over castes. Moreover their study has underlined the importance of behavioural characters in the study of phylogeny.

But what are the most important characteristics of social wasps, including the hover wasps that are the main subject of this book? Their colonies can be composed of a very low number of individuals or reach up to several hundreds of thousands. These societies live in nests made from various materials and built with a great variety of architectural solutions. The main function of the nest is to defend the growing larvae from predators, to furnish them with the most favourable climatic conditions and to serve as protection and support for the adults. Let us give a quick look at the Polistinae and Vespinae before speaking about the Stenogastrinae.

1.3.1 Polistinae

Among social wasps, those of the subfamily Polistinae exhibit various degrees of eusociality and a great adaptive radiation with 29 described genera all over the tropical and temperate areas of the world. Species belonging to the genus *Polistes* are the best studied from the behavioural point of view (Fig. 1.15). They are considered useful models to test sociobiological theories and in recent years they have been the subject of a large number of experimental studies (Turillazzi and West-Eberhard 1996; Starks and Turillazzi 2006). Their wide, almost global, geographical distribution has favoured their study, together with the limited size of their colonies, where the nest is composed of a single, non-enveloped, comb which permits easy observation of social interactions. *Polistes* are primitively eusocial and have post-imaginal caste differentiation with reproductive skew mainly determined by behavioural dominance, with the formation of dominance hierarchies (Pardi 1942, 1946). *Polistes dominula* is the species which has been

Fig. 1.14 A colony of *Liostenogaster flavolineata* (Stenogastrinae) (photo by David Baracchi)

studied the most and its recent introduction in North America (Cervo et al. 2000) means it has become a model species for American wasp students as well.

Nevertheless, with the exception of a few other species, totalling over 200 described so far, *Polistes* are still practically unknown from many points of view, mainly regarding variations in the ecology and natural history. Comparative studies on species living in more extreme habitats could allow us to perform interesting observations on the relationships between sociality and the environment in an apparently uniform genus. The genus *Polistes* also includes three particular species, restricted to the Mediterranean and Caspian basins, which are social parasites. Their fertilised females are not capable of building a nest of their own but must usurp that of another *Polistes* species where the larvae of their reproductives are then reared by the workers of the host species (Fig. 1.16) (Cervo and Dani 1996).

Similar to *Polistes* in social organisation, nest design and colony size are the more than 200 species belonging to the genus *Mischocyttarus* which live exclusively in the New World. Nest foundation can be associative or solitary (Jeanne 1972; Gadagkar 1991). Colonies of the genus *Parapolybia* also present characteristics similar to those of *Polistes* but are restricted to an East Asian

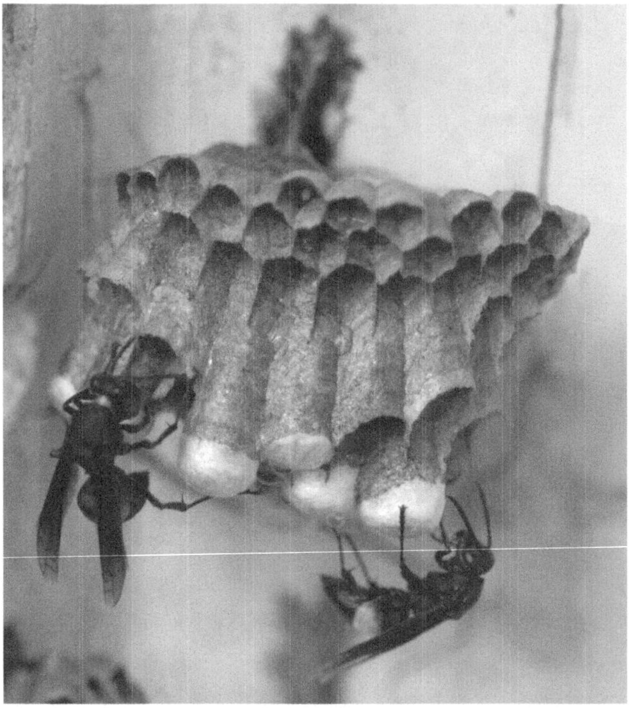

Fig. 1.15 Colony of *Polistes* sp. from Vietnam (photo by S. Bambi)

Fig. 1.16 Fight between a female of the social parasite *Polistes sulcifer* (*on the left*) and a female of the host species *Polistes dominula*

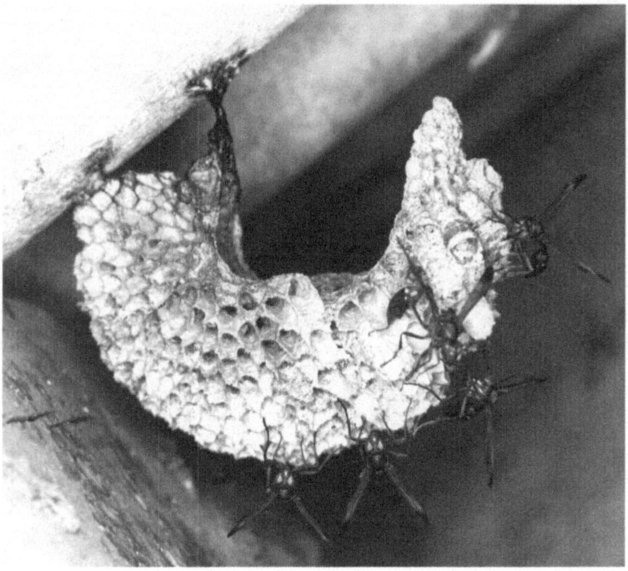

Fig. 1.17 *Belonogaster juncea*, a common species from Central Africa (photo by Riccardo Innocenti)

distribution (Gadagkar 1991). *Belonogaster*, instead, are African social wasps with only one species (*B. indica*) extending its distribution to South Arabia and India. These are quite interesting, usually large, elegant wasps (Fig. 1.17). Their nests, with no external envelopes, are of peculiar design and resemble an umbrella handle. The biology of a few species is known (Pardi 1977; Keeping 1992; Tindo et al. 1997).

All the genera I have mentioned so far are "independent founding" which means that colonies are initiated by a single or several fertilised and potentially reproductive females both in temperate and in tropical areas. *Parapolybia* and *Belonogaster* show another peculiarity with respect to *Polistes* and *Mischocyttarus*. Like the other genera belonging to the tribe Ropalidiini, the adults open a hole in the bottom of the cell where a larva is pupating to get rid of the larval faeces, then they reseal the bottom of the cell with salivary secretion which becomes a small, hard transparent window.

The swarm-founding Polistinae, by contrast, present larger and more organised colonies. In some species colony size reaches tens and hundreds of thousands of individuals: the largest colony ever reported was one belonging to the species *Agelaia vicina* collected in Brazil. This consisted of no less than one million three hundred thousand individuals (Zucchi et al. 1995). Their social organisation is complex with several queens which, in some species are morphologically distinct from the workers. In South Asia, only species of two genera of swarming Polistinae, both belonging to the tribe Ropalidiini, can be found: *Ropalidia* and *Polybioides*. The first is a very important genus for the study of social evolution as it includes

Fig. 1.18 Colonies of independent-founding (**a**) and swarm-founding *Ropalidia* (**b**)

both independent- and swarm-founding species (Fig. 1.18a, b) (Gadagkar 1991, 2001). Some independent-founding species, which are very common in Asia, Africa and Australia and prevalently inter-tropical in distribution, have been intensively studied by Japanese and Indian researchers (see Gadagkar 2001). They have, like *Polistes* wasps, very small colonies and non-enveloped nests. Other species of *Ropalidia* construct larger colonies and in some cases they protect their nests with envelopes which can be built of secretions produced by the wasps themselves (such as *R. opifex* from Malaysia; Maschwitz et al. 1990) or with paper. Differences between workers and queens are not all that striking in these wasps but in some species (such as *R. montana*) which have very large colonies, the two castes can be easily distinguished. Colonies of *Ropalidia* are quite common in a variety of ecosystems.

Polybioides is a genus with a limited number of species and has a quite peculiar geographical distribution as it occurs in both Equatorial Africa and the forests of South East Asia. Information about them is rather scarce. *P. raphigastra* is a very aggressive Asian wasp which nests in hollow trees or small cavities (van der Vecht 1966). Workers of this species can attack a person at considerable distance from the nest and continue to pursue their victims for several minutes. Their aggressiveness is so pronounced that they were used in booby traps built by the Vietcong during the Vietnam War (Spradbery 1973). The nest consists of a unique twirled comb which is surrounded by a pluri-layered envelope: a nest collected in West Malaysia exceeded 10,000 cells.

These wasps exhibit morphological differences between queens and workers (Turillazzi et al. 1994) and the study of their societies would undoubtedly produce interesting results, since they have developed social characteristics which have evolved independently but are quite similar to those of the Neotropical swarm-founding Polistinae. A much larger number of genera (22) of swarm founding

Polistinae live in South and Central America as far as southern USA.; here this subfamily has diversified in a great adaptive radiation. All these wasps are included in the tribe Epiponini and their success probably derives from the fact that even the phase in the biological cycle which in other wasps is the most critical, for them is social—i.e. foundation of new colonies. The colonial reproduction phase begins when new reproductive individuals, males and females, are produced and the next colonial cycle may take place in the same nest. In various genera, colonies with more than one queen are common, but in this case their reproductive potential is far lower than that of queens in the monogynic colonies of the Vespinae and independent-founding Polistinae. Again, the number of reproductive daughters these queens can generate is limited which lowers the number of new colonies that can be produced. On the other hand, colonies founded by swarming have a greater probability of surviving as a swarm of hundreds of wasps can build a nest in just a few hours which can then be easily defended from attack by ants and other predators.

Only some species in the vast range of these wasps have been studied, at least in part, for their social biology and natural history. One can find some primers on their biology in the excellent chapters by Bob Jeanne in the book by Ross and Matthews "The Social Biology of Wasps" which still represents the main reference for the study of these insects (Jeanne 1991a, b; Ross and Matthews 1991). Among the most studied, in any case, are species of the genus *Apoica*, pale nocturnal wasps which build nests without any envelope at all (Pickett and Wenzel 2007). Species belonging to *Agelaia*, with the larger colonies of all social wasps, in some cases forage on carrion, not a common behaviour in social wasps. Some species of the genus *Brachygastra* store large amounts of honey in their nests which are sought by natives in various localities in South America. These wasps, like honeybees, also exhibit autotomy of the sting apparatus (a trait also found in some species of *Polybioides*).

Metapolybia aztecoides (Fig. 1.19) was studied in depth by Mary Jane West-Eberhard who also observed a special "testing" behaviour carried out by workers towards young queens to ascertain which are the most reactive. This individual will then remain as the only queen in the colony while the others will be submitted or expelled from the nest (West-Eberhard 1978) (see also Nascimento et al. 2004 on *Asteloeca*, a close relative of *Metapolybia*). *Parachartergus apicalis*, called "Paco" by researchers, does not sting but sprays venom against its enemies, birds in particular (Jeanne and Keeping 1995). The genus *Polybia* includes more than 50 species; *P. emaciata* builds heavy nests, sometimes (though rarely) reaching the size of a football, made entirely from mud, while *P. occidentalis* is quite common, which the indigenous populations seek out to collect their honey. Large-sized wasps, up to 2.5 cm in length, are those belonging to the genus *Synoeca* which, like some *Parachartergus* species, build nests as long as 1.5 m on tree trunks.

Protopolybia sedula has been studied by Naumann (1970, 1975). These wasps can swarm several times a year and present a well-marked division of labour between the workers, the youngest working in the nest and the older ones foraging. Foragers, on their return to the nest, perform ritualised dances before leaving again,

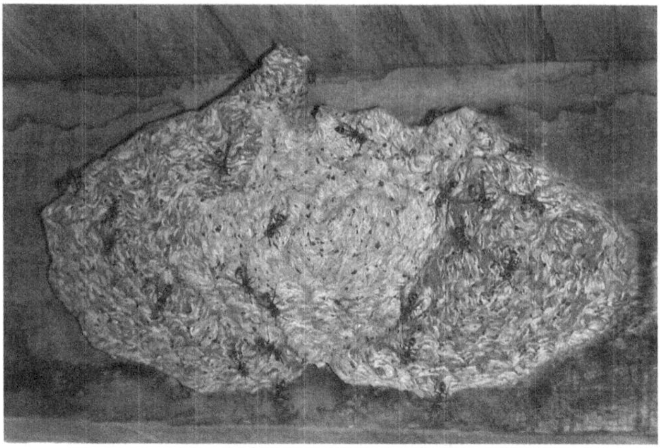

Fig. 1.19 A colony of *Metapolybia atzecoides*. (Courtesy James Carpenter)

suggesting some kind of communication regarding the food source (Naumann 1970). At present, however, no evidence for recruitment to a specific foraging site has been collected for any swarming polistine wasp (Raveret Richter 2000). But species of the large-colony polybiine genera do possess efficient chemical communication during swarming movements (reviewed in Smith et al. 2002) to new nest sites.

Swarm-founding polistine wasps also use pheromones, which are usually components of the venom secretion, for alarm and for coordinating colony defence against predators (Kojima 1994; Fortunato et al. 2004 on Ropalidiini; Jeanne 1981; O'Donnell et al. 1997; Dani et al. 2000 on Epiponini). This capacity is also present in some independent-founding species (Bruschini et al. 2006).

The nests of these wasps offer a wide variety of architectural designs (Wenzel 1991); I shall return to this subject in the section on hover wasp nests.

1.3.2 Vespinae

There are only four genera in the subfamily Vespinae: *Provespa, Vespa, Vespula* and *Dolichovespula* (Matsuura 1991; Greene 1991). All of them build nests which are characteristically formed of stacked combs surrounded by mono or multi-stratified envelopes (Fig. 1.20). *Provespa* occurs exclusively in South East Asia and includes only nocturnal species; males and females of these wasps fly at night and are easily attracted to lamps or by lights at the forest edge. These wasps also present swarming nest foundation (Matsuura and Yamane 1990).

Species of the genus *Vespa* are distributed only in the old world (with the exception of *Vespa crabro* imported in the USA). These wasps, popularly called hornets, are among the largest of living social insects. *Vespa mandarinia* is the

Fig 1.20 A colony of *Vespa crabro* (photo by Elisabetta Francescato)

largest social wasp (and certainly the largest social insect if we exclude the physogastric queens of some termites) and its queens reach 5.5 cm in length! Species of hornets often attack other social wasps and in various countries they are considered as apiculture pests for they can destroy entire honeybee colonies in a very short time. As we shall see, these wasps have probably deeply influenced the social evolution, distribution and nest architecture of the Stenogastrinae and other social wasps in South East Asia. Vespine wasps exhibit clear morphological differences between queens and workers, large size colonies and complex social organisations which are organised through the use of pheromones (Matsuura 1991).

Vespula and *Dolichovespula* include species which are present in the temperate regions of the old world and North America. *Vespula germanica* and *V. vulgaris* form quite large colonies and usually build their nests in underground cavities. They are scavengers and this explains the quite frequent contacts with man. Recently these species have been introduced through human activities to various extra-areal zones such as Chile, Australia and New Zealand causing great problems (Chapman and Bourke 2008). Various books can be used as reference for an in-deep survey of their social biology: the ones from Spradbery (1973) and Edwards (1980) remain very good primers while a more recent review is the book of Matsuura and Yamane (1990).

References

Baracchi D, Dapporto L, Teseo S, Hashim R, Turillazzi S (2010) Medium molecular weight polar substances of the cuticle as tools in the study of the taxonomy, systematics and chemical ecology of tropical hover wasps (Hymenoptera: Stenogastrinae). J Zool Syst Evol Res 48:109–114

Baracchi D, Dapporto L, Turillazzi S (2011) Relevance of wing morphology in distinguishing and classifying genera and species of Stenogastrinae wasps. Contrib Zool 80:191–199

Bruschini C, Cervo R, Turillazzi S (2006) Evidence of alarm pheromones in the venom of *Polistes dominulus* workers (Hymenoptera: Vespidae). Physiol Entomol 31:286–293

Carpenter JM (1982) The phylogenetic relationships and natural classification of the Vespoidea (Hymenoptera). Syst Entomol 7:11–38

Carpenter JM (1988) The phylogenetic system of the Stenogastrinae (Hymenoptera: Vespidae). J N Y Entomol Soc 96:140–175

Carpenter JM (1991) Phylogenetic relationships and the origin of social behavior in the Vespidae. In: Ross KG, Matthews RW (eds) The social biology of wasps. Cornell University Press, Ithaca, NY

Carpenter JM (2001) New generic synonymy in Stenogastrinae (Insecta: Hymenoptera; Vespidae). Nat Hist Bull Ibaraki Univ 5:27–30

Carpenter JM (2003) On "Molecular phylogeny of Vespidae (Hymenoptera) and the evolution of sociality in Wasps. Am Mus Novit 3389:1–20

Carpenter JM, Kojima J (1996) Checklist of the subfamily Stenogastrinae (Hymenoptera: Vespidae). J N Y Entomol Soc 104:21–36

Carpenter JM, Kojima J (2006) Checklist of the species of the subfamily Stenogastrinae Bequaert, 1918 (Hymenoptera: Vespidae). Natural History Laboratory, Ibaraki University (IUNH) and World Association for the Study of Paperwasps (WASP), Japan. 13:5–26

Carpenter JM, Starr CK (2000) A new genus of hover wasps from Southeast Asia (Hymenoptera: Vespidae; Stenogastrinae). Am Mus Novit 3291:1–12

Cervo R, Dani FR (1996) Social parasitism and its evolution in Polistes. In: Turillazzi S, West-Eberhad MJ (eds) Natural history and evolution of paper-wasps. Oxford University Press, Oxford

Cervo R, Zacchi F, Turillazzi S (2000) *Polistes dominulus* (Hymenoptera: Vespidae) invading North America: some hypotheses for its rapid spread. Insect Soc 47:155–157

Chapman RE, Bourke AFG (2008) The influence of sociality on the conservation biology of social insects. Ecol Lett 4:650–662

Dani FR, Jeanne RL, Clarke SR, Jones GR, Morgan ED, Francke W, Turillazzi S (2000) Chemical characterization of the alarm pheromone in the venom of *Polybia occidentalis* and of volatiles from the venom of *P. sericea*. Physiol Entomol 25:363–369

Das BP, Gupta VK (1984) A catalogue of the families Stenogastridae and Vespidae from the Indian subregion (Hymenoptera: Vespidae). Orient Insects 17:395–464

De Saussure H (1852–1858) Monographie Des Guêpes Sociales Ou de la Tribu Des Vespiens. Masson/J. Cherbuliez, Paris/Genève

Dong D, Otsuka K (1997) The Stenogastridae (Hymenoptera: Vespoidea) of China. J Southwestern Agric Univ 19:205–212 (In Chinese)

Dover C, Rao R (1922) A note on the diplopterous wasps in the collection of the Indian Museum. J Asiat Soc Bengal 18:235–249

Edwards R (1980) Social wasps, their biology and control. Rentokil Limited, East Grinstead

Evans HE, West-Eberhard MJ (1970) The wasps. University Michigan Press, Ann Arbor, MI

Fortunato A, Dani FR, Sledge MF, Fondelli L, Turillazzi S (2004) Alarm communication in *Ropalidia* social wasps. Insect Soc 51:299–305

Gadagkar R (1991) *Belonogaster*, *Mischocyttarus*, *Parapolybia*, and independent founding *Ropalidia*. In: Ross KG, Matthews RW (eds) The social biology of wasps. Cornell University Press, Ithaca

Gadagkar R (2001) The social biology of *Ropalidia marginata*: toward understanding the evolution of eusociality. Harvard University Press, Cambridge, MA

Greene A (1991) *Dolichovespula* and *Vespula*. In: Ross KG, Matthews RW (eds) The social biology of wasps. Cornell University Press, Ithaca, NY

Guérin-Méneville FE (1831) Plates from Voyage autour du monde sur la Coquille. In: Duperrey LI (ed) Voyage autour du monde, exécuté par ordre du roi, sur la corvette de sa majesté, La Coquille, pendant les années 1822, 1823, 1824 et 1825. H. Bertrand, Paris

Guérin-Méneville FE (1838) Crustacées, arachnides et insectes. In: Duperrey LI (ed) Voyage autour du monde, exécuté par ordre du roi, sur la corvette de sa majesté, La Coquille, pendant les années 1822, 1823, 1824 et 1825. H. Bertrand, Paris

Hines HM, Hunt JH, O'Connor TK, Gillespie JJ, Cameron SA (2007) Multigene phylogeny reveals eusociality evolved twice in vespid wasps. Proc Natl Acad Sci USA 104:3295–3299

Hunt JH (2007) The evolution of social wasps. Oxford University Press, New York

Hunt JH (2011) A conceptual model for the origin of worker behaviour and adaptation of eusociality. J Evol Biol 25:1–19

Ito Y (1993) Behaviour and social evolution of wasps: the communal aggregation hypothesis, Oxford series in ecology and evolution. Oxford University Press, Oxford

Iwata K (1967) Report of the fundamental research on the biological control of insect pests in Thailand. II. The report on the bionomics of Aculeate wasps bionomics of subsocial wasps of Stenogastrinae (Hymenoptera, Vespidae). Nat Life SEA 5:259–293

Iwata K (1971) Evolution of instinct. Comparative ethology of Hymenoptera. Mono-shoten, Kanagawa-ken. [in Japanese]. Smithsonian Institution and the National Science Foundation, Washington DC

Jacobson E (1935) Aantekeningen over Stenogastrime (Hym. Vespid.). Ent Med Soc 14:309–318

Jeanne RL (1972) Social biology of the neotropical wasp *Mischocyttarus drewseni*. Bull Mus Comp Zool 144:63–150

Jeanne RL (1980) Evolution of social behavior in the Vespidae. Annu Rev Entomol 25:371–396

Jeanne RL (1981) Alarm recruitment, attack behaviour, and the role of alarm pheromones in *Polybia occidentalis* (Hymenoptera: Vespidae). Behav Ecol Sociobiol 9:143–148

Jeanne RL (1991a) The swarm-founding Polistinae. In: Ross KG, Matthews RW (eds) The social biology of wasps. Cornell University Press, Ithaca, NY

Jeanne RL (1991b) Polyethism. In: Ross KG, Matthews RW (eds) The social biology of wasps. Cornell University Press, Ithaca, NY

Jeanne RL, Keeping MG (1995) Venom spraying in *Parachartergus colobopterus*: a novel defensive behavior in a social wasp (Hymenoptera: Vespidae). J Insect Behav 8:433–442

Keeping MG (1992) Social organization and division of labour in colonies of the polistine wasp, *Belonogaster petiolata*. Behav Ecol Sociobiol 31:211–224

Kojima J (1990) Immatures of hover wasps (Hymenoptera, Vespidae, Stenogastrinae). Jpn J Entomol 58:506–522

Kojima J (1994) Evidence for an alarm pheromone in the Old World, swarm-founding paper wasp, *Ropalidia romandi* (Le Guillou) (Hymenoptera: Vespidae). J Aust Entomol Soc 33:45–47

Krombein KV (1976) *Eustenogaster*, a primitive social Sinhalese wasp. Loris 14:303–306

Landi M, Queller DC, Turillazzi S, Strassmann JE (2003) Low relatedness and frequent queen turnover in the stenogastrine wasp *Eustenogaster fraterna* favor the life insurance over the haplodiploid hypo-thesis for the origin of eusociality. Insect Soc 50:262–267

Marino-Piccioli MT, Pardi L (1970) Studies on the biology of *Belonogaster* (Hymenoptera, Vespidae). On the behaviour pattern of *Belonogaster griseus* (Fab.). Monit Zool Ital 3:197–225

Marino-Piccioli MT, Pardi L (1978) Studies on the biology of *Belonogaster* (Hymenoptera: Vespidae): 3. The nest of *Belonogaster griseus*. (Fab.). Monit Zool Ital (NS) 10:179–228

Maschwitz U, Dorow WHO, Botz T (1990) Chemical composition of the nest walls, and nesting behaviour, of *Ropalidia* (*Icarielia*) *opifex* van der Vecht, 1962 (Hymenoptera: Vespidae), a Southeast Asian social wasp with translucent nests. J Nat Hist 24:1311–1319

Matsuura M (1991) *Vespa* and *Provespa*. In: Ross KG, Matthews RW (eds) The social biology of wasps. Cornell University Press, Ithaca, NY
Matsuura M, Yamane SK (1990) Biology of the vespine wasps. Springer, Berlin
Nascimento FS, Tannure-Nascimento IC, Zucchi R (2004) Behavioral mediators of the cyclical oligogyny in the Amazonian swarm-founding wasp Asteloeca ujhelyii (Hymenoptera, Vespidae, Epiponini). Insect Soc 51:17–23
Naumann MG (1970) The nesting behavior of Protopolybia pumila in Panama (Hymenoptera, Vespidae). PhD thesis, University of Kansas. Kansas
Naumann MG (1975) Swarming behavior: evidence for communication in social wasps. Science 189:642–644
O'Donnell S, Hunt JH, Jeanne RL (1997) Gaster-flagging during colony defense in neotropical swarm-founding wasps (Hymenoptera: Vespidae, Epiponini). J Kansas Entomol Soc 70:175–180
Pagden HT (1958) Some Malayan social wasps. Malayan Nat J 12:131–148
Pagden HT (1962) More about *Stenogaster*. Malayan Nat J 16:95–102
Pardi L (1942) Ricerche sui Polistini. V. La poliginia iniziale in *Polistes gallicus* (L.). Boll Ist Entom Univ Bologna 14:1–106
Pardi L (1946) Ricerche sui Polistini VII. La "dominazione" ed il ciclo ovarico annuale di *Polistes gallicus* (L.). Boll Ist Entom Univ Bologna 15:25–84
Pardi L (1977) Su alcuni aspetti della biologia di *Belonogaster* (Hymenoptera, Vespidae). Boll Ist Entom Univ Bologna 33:281–299
Pardi L (1980) Le vespe sociali: biologia ed evoluzione del comportamento. Accad Naz Lincei Contrib Centro Linceo Interdisc Sci Mat Appl 51:161–221
Pardi L, Marino-Piccioli MT (1970) Studies on the biology of *Belonogaster* (Hymenoptera, Vespidae). 2. Incipient caste differentiation in *B.griseus* (Fab.). Monit Zool Ital 3:235–265
Pardi L, Marino-Piccioli MT (1981) Studies on the biology of *Belonogaster* (Hymenoptera, Vespidae): 4. On caste differences in Belonogaster griseus and the position of this genus among social wasps. Monit Zool Ital 14:131–146
Pardi L, Turillazzi S (1982) Biologia delle Stenogastrinae. Atti Acc Naz Ital Ent Rdc XXX:3–21
Pickett KM, Carpenter JM (2010) Simultaneous analysis and the origin of eusociality in the Vespidae (Insecta: Hymenoptera). Arth Syst Phyl 68:3–33
Pickett KM, Wenzel JW (2007) Revision and cladistic analysis of the nocturnal social wasp genus, *Apoica* Lepeletier (Hymenoptera, Vespidae, Polistinae, Epiponini). Am Mus Nov 3562:1–30
Raveret Richter M (2000) Social wasps (Hymenoptera: Vespidae) foraging behaviour. Annu Rev Entomol 45:121–150
Richards OW (1971) The biology of the social wasps (Hymenoptera, Vespidae). Biol Rev 46:483–528
Ross KG, Matthews RW (1991) The social biology of wasps. Cornell University Press, Ithaca, NY
Saito F, Kojima JL (2007) A taxonomic revision of the hover wasp genus *Eustenogaster* van der Vecht (Insecta: Hymenoptera; Vespidae, Stenogastrinae). Zootaxa 1556:1–30
Saito F, Nguyen LPT, Carpenter JM, Kojima J (2006) A new *Eustenogaster* species (Hymenoptera: Vespidae; Stenogastrinae), the first hover wasp known to overwinter on the nest. Am Mus Nov 3534:1–12
Saito F, Nguyen LPT, Kojima J (2009) Colony cycle of a "temperate" hover wasp, *Eustenogaster nigra*, with special reference to overwintering of males in an enveloped nest together with virgin females (Hymenoptera, Vespidae, Stenogastrinae). Insect Soc 56:49–54
Samuel C (1987) Factors affecting colony size in the stenogastrine wasp *Liostenogaster flavolineata*. PhD thesis, University of Malaya, Kuala Lumpur
Schmitz J, Moritz RFA (1998) Molecular phylogeny of Vespidae (Hymenoptera) and the evolution of sociality in wasps. Mol Phylogenet Evol 9:183–191
Smith AR, O'Donnell S, Jeanne RL (2002) Evolution of swarm communication in eusocial wasps (Hymenoptera: Vespidae). J Insect Behav 15:751–764

Spradbery JP (1973) Wasps: an account of the biology and natural history of solitary and social wasps. Sidgwick and Jackson/University of Washington Press, London/Seattle
Spradbery JP (1975) The biology of *Stenogaster concinna* Van der Vecht with comments on the phylogeny of Stenogastrinae (Hymenoptera: Vespidae). J Aust Entomol Soc 14:309–318
Spradbery JP (1989) The nesting of *Anischnogaster iridipennis* (Smith) (Hymenoptera: Vespidae) in New Guinea. J Aust Entomol Soc 28:225–228
Starks PT, Turillazzi S (2006) *Polistes* paper wasps: emergence of a model genus. Ann Zool Fenn 43:385–386
Starr CK (1984) Rambling remarks on Philippine wasp-mimics. Sphecos 9:11–13
Strassmann J, Hughes CR, Turillazzi S, Solis CR, Queller DC (1994) Genetic relatedness and incipient eusociality in stenogastrine wasps. Anim Behav 48:813–821
Tindo M, D'Agostino P, Francescato E, Dejean A, Turillazzi S (1997) Associative colony foundation in the tropical wasp *Belonogaster juncea juncea* (Hymenoptera: Vespidae). Insect Soc 44:365–377
Turillazzi S (1988) Adults and nest of *Liostenogaster vechti* n.sp. (Hymenoptera Stenogastrinae). Trop Zool 1:193–201
Turillazzi S (1999) New species of *Liostenogaster* van der Vecht 1969, with keys to adults and nests (Hymenoptera Vespidae Stenogastrinae). Trop Zool 12:335–358
Turillazzi S, Carfi S (1996) Adults and nest of *Liostenogaster pardii* n.sp. (Hymenoptera Stenogastrinae). Trop Zool 9:19–30
Turillazzi S, West-Eberhard MJ (1996) Natural history and evolution of paper-wasps. Oxford University Press, Oxford
Turillazzi S, Francescato E, Baldini Tosi A, Carpenter JM (1994) A distinct caste difference in *Polybioides tabhidus* (Fabricius) (Hymenoptera: Vespidae). Insect Soc 41:327–330
van der Vecht J (1966) The East Asiatic and Indo-Australian species of *Polybioides* du Buysson and *Parapolybia* de Saussure (Hym., Vespidae). Zool Verh 82:1–42
van der Vecht J (1972) A review of the new genus *Anischnogaster* in the Papua region (Hymenoptera, Vespidae). Zool Meded Leiden 47:240–256
van der Vecht J (1975) A review of the genus *Stenogaster* Guerin (Hymenoptera: Vespoidea). J Aust Entomol Soc 14:283–308
van der Vecht J (1977) Studies of oriental Stenogastrinae (Hymenoptera: Vespoidea). Tijdschr Entomol 120:55–75
von Schulthess A (1927) Fauna Sumatrensis. Vespidae (Hym). Suppl Entomol 16:81–92
Wenzel JW (1991) Evolution of nest architecture. In: Ross KG, Matthews RW (eds) The social biology of wasps. Cornell University Press, Ithaca, NY
West-Eberhard MJ (1978) Temporary queens in *Metapolybia* wasps: non-reproductive helpers without altruism? Science 200:441–443
Williams FX (1919) Philippine wasp studies, part 2: descriptions of new species and life history studies. Bull Exp Sta Hawaii Sugar Planters' Assoc Entomol 14:19–186
Williams FX (1928) Studies in tropical wasps their hosts and associates (with descriptions of new species). Bull Exp Sta Hawaii Sugar Planters' Assoc Entomol 19:1–179
Wilson EO (1971) The insect societies. Belknap, Harvard University Press, Cambridge, MA
Yoshikawa K, Ohgushi R, Sakagami SF (1969) Preliminary report on entomology of the Osaka City University 5th Scientific Expedition to Southeast Asia 1966. With description of two new genera of stenogasterine [sic] wasps by J. van der Vecht. Nat Lif SEA 6:153–182
Zucchi R, Sakagami SF, Noll FB, Mechi MR, Mateus S, Baio MV (1995) *Agelaia vicina*, a swarm-founding polistine with the largest colony size among wasps and bees (Hymenoptera: Vespidae). J N Y Entomol Soc 103:129–137

Chapter 2
Morphology and Anatomy

2.1 Morphology and Anatomy of the Adults

Hover wasps are the most beautiful of all social wasps. Most of them are not really colourful, but their elegance, especially when flying, competes against that of better known and more loved insects.

They can be considered medium-size wasps and their length varies between 10 mm (some species of *Parischnogaster)* and 25 mm (some *Eustenogaster*).

The colour in most genera is dark brown with brilliant shades of black mixed with yellow, white or yellowish spots, but in *Liostenogaster* both sexes are brown with more or less extended yellow areas. Some species present distinct black markings on the face that could be important for individual recognition (Fig. 2.2). The wings, transparent, have iridescent reflections in some species.

The body (Fig. 2.1) is distinctly divided into three parts: the head, with the antennae and mouthparts, the thorax (with the legs and wings) plus the first abdominal segment (the propodeum), and the gaster the first segment of which is a long petiole, truncate posteriorly. For a general description I shall refer to the scheme elaborated by Spradbery (1973) for vespine wasps, stressing the differences when necessary, and taking examples from species in the various genera. To summarise the main characteristics of some body parts in the various genera, I shall also use the technique known as TPS (thin-plate spline graphical analysis, Rohlf 2006, 2007), which compares landmark points of the exoskeleton obtained through specific algorithms, resulting in schematisation of the variability in a given sample population.

The matrix in Table 2.1, developed by Carpenter (2001) for the phylogenetic analysis of the genera summarises the principal characteristics of the group.

S. Turillazzi, *The Biology of Hover Wasps*, DOI 10.1007/978-3-642-32680-6_2,

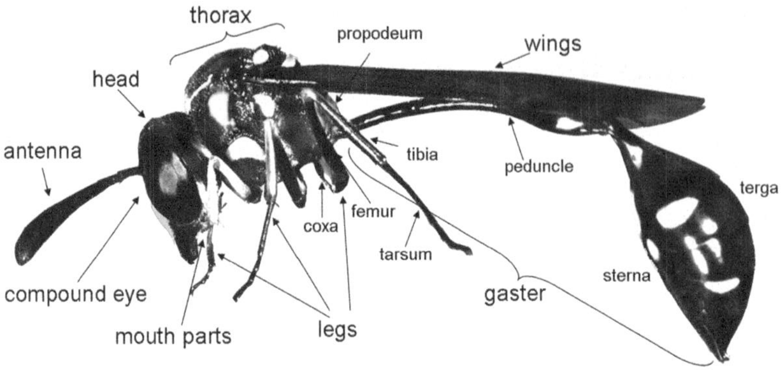

Fig. 2.1 Schematic picture of a female hover wasp (*Parischnogaster* sp.), indicating the main morphological parts

Table 2.1 Data on genera of Stenogastrinae (from Carpenter 2001)

Genera	Lio	Par	Met	Ani	Ste	Eus	Coc
Occip. carina	0	1	1	1	1	1	1
Male clypeus	0	0	0	0	2	1	1
Male teeth	$	0	1	0	2	2	1
Maxillary palp	0	0	0	0	1	1	*
Male antennae	0	0	1	0	2	0	3
Notauli	0	0	1	0	0	0	1
Scutellum	0	0	0	0	0	1	*
Prop. valvula	0	1	2	0	0	0	1
Prop. sculpture	$	$	0	1	0	2	2
II metasomal	0	1	1	1	0	0	0
Parameral spine	0	1	0	0	0	0	2
Aedeagus	0	1	1	0	0	0	1
Aedeagal Processes	0	0	1	0	0	0	1

1. Occipital carina: gap (0); fused to hypostomal carina (1)
2. Male clypeus: pointed apically (0); round (1); depressed (2); emarginate (3)
3. Male mandibular teeth: three (0); two (1); one (2); four or five (3)
4. Maxillary palpi: palpomeres 2 and 3 equal in length (0); 2 longer than 3 (1)
5. Male antennae: conical (0); flat (1); tipped (2); spatulate (3)
6. Notauli: weak (0); strong (1)
7. Scutellum: ecarinate (0); carinate in female (1)
8. Propodeal valvula: round (0); posteriorly attenuate (1); narrow (2)
9. Propodeal sculpture: striate (0); punctate (1); smooth (2)
10. Metasomal segment II: not petiolate (0); petiolate (1)
11. Parameral spine: spinose (0); flat (1); elbowed (2)
12. Aedeagus: normal (0); dilated (1)
13. Aedeagal processes: absent (0); present laterobasally (1)

An *asterisk* (*) denotes a polymorphism showing all applicable states; a *dollar sign* ($) denotes a subset polymorphism (for *Liostenogaster*: Male mandibular teeth [0,1] and Propodeal structure [1,2]; for *Parischnogaster*: Propodeal structure [0,1]). Multistate characters are treated as non-additive

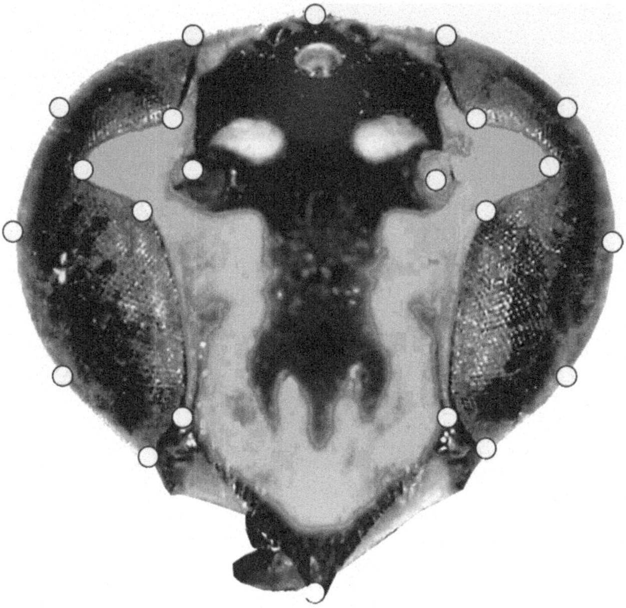

Fig. 2.2 Head of female *Liostenogaster flavolineata,* indicating the landmark points used for the TPS analysis

2.1.1 The Head

The shape of the head, viewed from the front, is sub-triangular owing to the very long mandibles (Fig. 2.2). This is particularly evident in *Stenogaster* (with average head length/width ratio of over 1).

According to TPS analysis performed on 26 landmark points on the head of females of 27 species from 6 genera, the largest variation in species occurs in head width rather than head length (Fig. 2.3). On the right side of the graph we find *Eustenogaster* and *Stenogaster* with heads longer than wide, while *Parischnogaster* presents the widest and shortest shape and *Liostenogaster* and *Metischnogaster* something in between (Fig. 2.3).

The clypeus is usually quite pointed in females and slightly less in the males, with the exception of males of *Eustenogaster* and *Metischnogaster,* which have a rounded clypeus, and *Stenogaster* where it is rounded apically and slightly depressed.

A particular carina surrounding the occipital foramen (the occipital carina) fuses with the hypostomal carina in all the genera except *Liostenogaster*. This represents a distinction for this genus (Fig. 2.4).

The female antennae are generally somewhat clavate. In males they are generally conical but in *Metischnogaster* they are flattened, in *Stenogaster* truncate and in

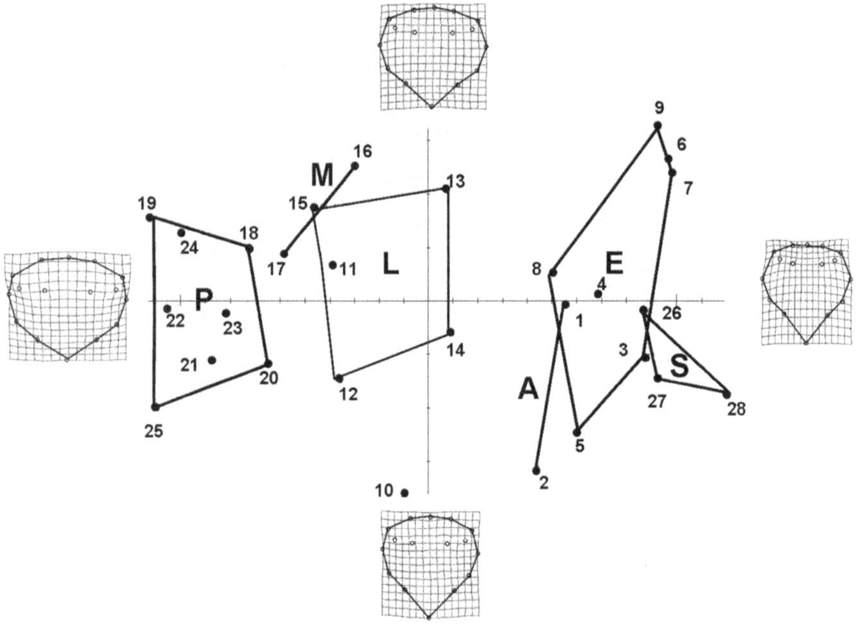

Fig. 2.3 General graph of head shape in 28 species belonging to 6 genera. obtained by TPS analysis of the landmark points indicated in Fig. 2.2. A = *Anischnogaster* (1 = *A. iridipennis*, 2 = *A. spilaspis*). E = *Eustenogaster* (3 = *E. latebricola*, 4 = *E. calyptodoma*, 5 = *E. fumipennis*, 6 = *E. hauxwellii*, 7 = *E. luzonensis*, 8 = *E. micans*, 9 = *E. palavanica*). L = *Liostenogaster* (11 = *L. flavolineata*, 12 = *L. nitidipennis*, 13 = *L. topographica*, 14 = *L. varipicta*, 15 = *L. vechti*). M = *Metischnogaster* (16 = *M. cilipennis*, 17 = *M. drewseni*). P = *Parischnogaster* (18 = *P.alternata*, 19 = *P. depressigaster*, 20 = *P. jacobsoni*, 21 = *P. mellyi*, 22 = *P. nigricans*, 23 = *P. striatula*, 24 = *P. timida*, 25 = *P. unicuspata*). S = *Stenogaster* (26 = *S. canaliculata*, 27 = *S. concinna*, 28 = *S. latebricola*). 10 = *Holischnogaster* (=*Parischnogaster*) *gracilipes*)

the males of the genus *Cochlischnogaster* there is a final spatulate segment with a peculiar spoon shape (Fig. 2.5).

Regarding the mouthparts, the mandibles are much thinner than those in polistine or vespine wasps: in the females they can bear up to three teeth, which also occur in the males of *Parischnogaster* and *Anischnogaster*. The mandibles of the males of *Metischnogaster* and *Cochlischnogaster* have two teeth, but in *Eustenogaster* and *Stenogaster* there is only one with the internal margin almost straight. The second tooth of the female mandible can be blunt or sharply edged depending on the material the species uses for nest construction (see Chap. 6, Fig. 6.72).

Eustenogaster and *Stenogaster* have a second maxillary palp which is three times longer than the third; the palpomeres are almost the same length as in the other genera.

The compound eyes are large; in male *Cochlischnogaster*, for example, Carpenter and Starr (2000) calculate that they are about 85 % as long as the entire head,

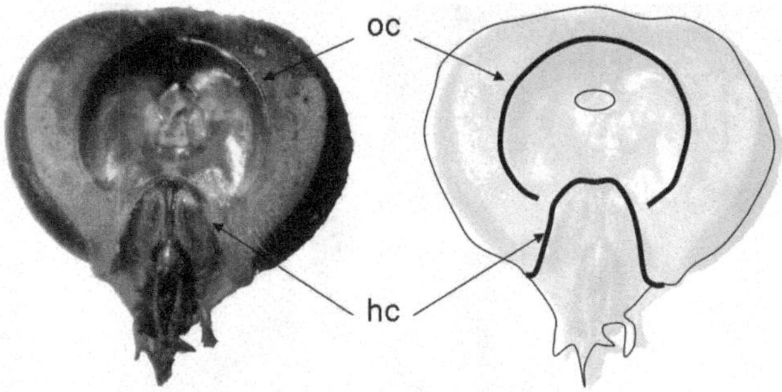

Fig. 2.4 Occipital view of the head of a *Liostenogaster* showing the occipital carina (oc) not reaching the hypostomal carina (hc)

Fig. 2.5 Antennae of a male of *Cochlishnogaster* with the peculiar final segment (courtesy of James Carpenter)

occupying almost all the side view of the head itself. The ocelli are also large if compared with those of most polistine and vespine wasps (personal observations), this could be related to the rather dark environments where the hover wasps live.

The colouration of the "face" is particularly evident in females and males of some species (*Liostenogaster*, *Eustenogaster*, *Metischnogaster* and some species of *Parischnogaster*) with widely varying intraspecific (and even intracolonial) patterning, while in other species (such as *P. mellyi* and *P. nigricans serrei*) facial colouration is quite uniform. Males usually have less evident facial markings than females (Fig. 2.6).

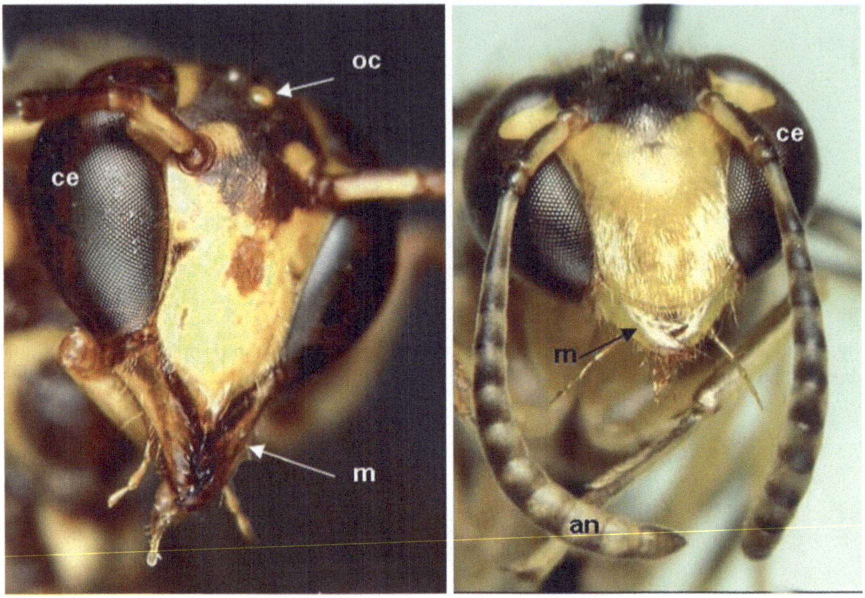

Fig. 2.6 Heads of female and male of *Liostenogaster topographica* with different facial markings (*oc* ocellum; *ce* compound eye; *m* mandible, *an* antenna)

2.1.2 The Thorax

The thorax is massive and globular (Fig. 2.7). The notauli (paired lines or grooves on the mesoscutum that subdivide the sclerite into a median midlobe and lateral lobes) are weak with the exception of *Metischnogaster* and *Cochlischnogaster*. The scutellum lacks a carina in all the genera with the exception of females of *Eustenogaster* and *Cochlischnogaster*.

The propodeum, namely the first abdominal segment, is attached to the thorax as in the other aculeate Hymenoptera. The cuticular sculpture of this segment can be striate (as in *Liostenogaster*, *Anischnogaster*, *Stenogaster* and *Eustenogaster*), punctate (as in *Parischnogaster* and *Cochlischnogaster*) or smooth as in *Metischnogaster*.

The legs are short and delicate compared to those of other social wasps. The fore and hind tibiae present comb-like appendages (calcars) which, together with the proximal part of the first tarsal segments, serve to clean the antennae and wings respectively. In males of *Anischnogaster loriai* a special structure (metatarsal organ) that van der Vecht (1972) considers to be connected to a glandular apparatus with unspecified function, is associated with the calcar of the hind legs.

The fore and hind wings differ in size (Fig. 2.8) and the fore wings are not folded longitudinally at rest as in polistine and vespine wasps. This character does not seem to be related to the evolved social biology of the two subfamilies (for example to avoid these insects being encumbered by their wings when they are on the nest) because this peculiar feature is also present in the Eumeninae and some Masarinae.

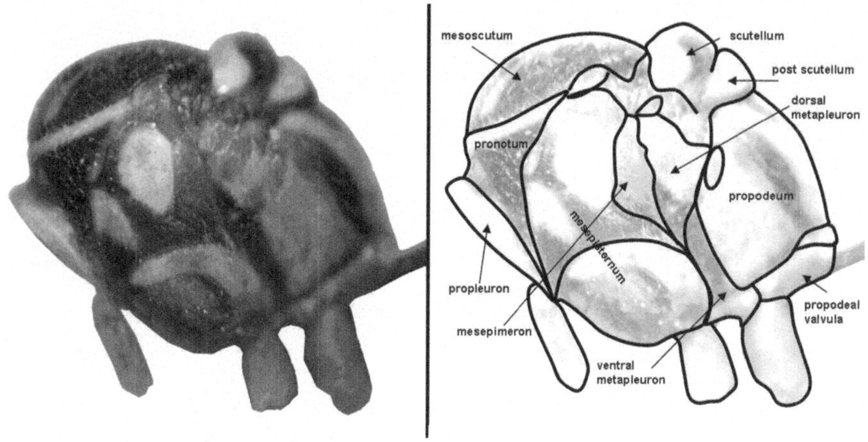

Fig. 2.7 Lateral view of the thorax of *Liostenogaster flavolineata* with a scheme of the component parts

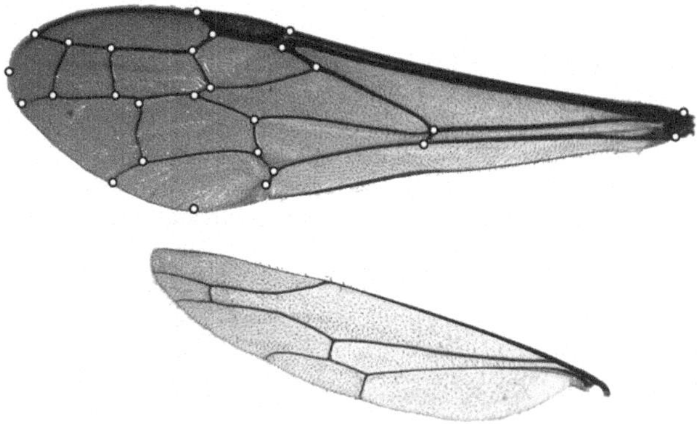

Fig. 2.8 Anterior and posterior wings of *Liostenogaster vechti*. Note the landmark points on the fore wing used for TPS analysis in Fig. 2.9

Variation among genera is small; the general shape of the male anterior wings is summarised in the TPS graph in Fig. 2.9 obtained from the analysis of 27 landmark points from one specimen of 16 species belonging to 6 different genera (Hg: *Holischnogaster* (=*Parischnogaster*) *gracilipes*). Species of the genera *Liostenogaster* and *Eustenogaster* exhibit more distinctive shapes while *Parischnogaster*, *Anischnogaster* and *Stenogaster* share a similar structure. This could be related to the habit the males of these genera have in performing aerial patrol during particular times of the day, hovering and protecting precise perching sites. We shall come back to this argument in other parts of the book.

When in flight, the fore and hind wings are coupled together as the anterior margin of the hind wings bears a row of small hooks (hamuli) which catch onto a particular vein of the fore wings. The number of these hamuli ranges from a

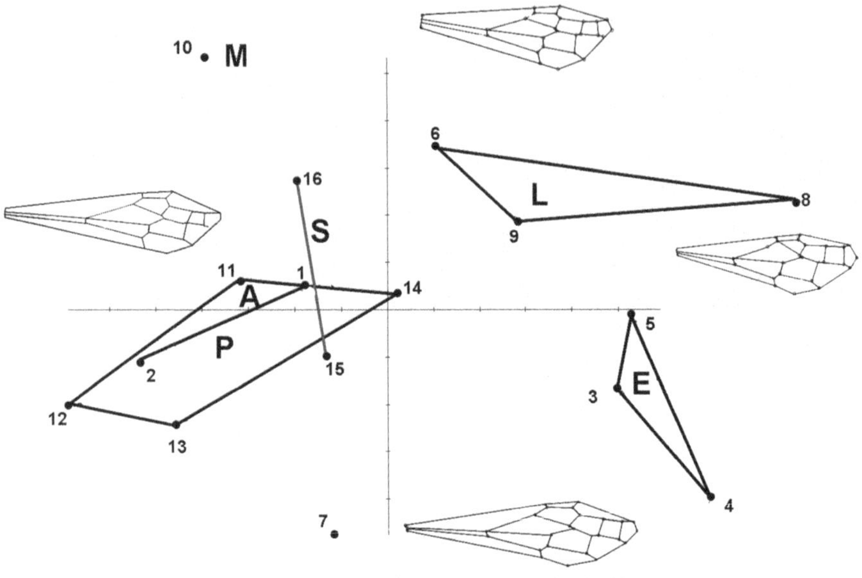

Fig. 2.9 General graph of the right fore wing shape of 16 species belonging to 6 genera obtained with TPS analysis of the landmark points given in Fig. 2.8. A = *Anischnogaster* (1 = *A. laticeps*, 2 = *A. spilaspis*). E = *Eustenogaster* (3 = *E. calyptodoma*, 4 = *E. fraterna*, 5 = *E. micans*). L = *Liostenogaster* (6 = *L. flavolineata*, 8 = *L. pardii*, 9 = *L. vechti*). M = *Metischnogaster* (10 = *M. drewseni*). P = *Parischnogaster* (11 = *P. nigricans*,12 = P. *alternata*, 13 = *P. jacobsoni*, 14 = *P. mellyi*). S = *Stenogaster* (15 = *S. concinna*, 16 = *S. adusta*). (7 = *Holischnogaster* = *Parischnogaster gracilipes*)

minimum of 6, which I counted in a female of *Metischnogaster drewseni*, to a maximum of 13, which I found in a male *Eustenogaster fraterna,* while there is no difference between males and females of the same species. The hind wings of male *Metischnogaster cilipennis* are characterised by a conspicuous fringe of hairs at the back. Wings are transparent but some species have iridescent reflections.

The flight of these wasps is characteristic and some genera in particular (i.e. *Parischnogaster*, *Metischnogaster* and *Eustenogaster*) can hover in the air remaining almost immobile, like small dragonflies. This is particularly efficient for their special way of foraging: females are able to fly close to spider webs from which they steal small prey (see Chap. 3, Fig. 3.10) (videoclip no.1) and, in some cases, bits of silk for nest construction.

2.1.3 The Gaster

The gaster is formed of the abdominal segments minus the first (propodeum) which is attached to the last segment of the thorax. The first segment of the gaster (metasomal segment 2) is called the "petiole" and in the hover wasps it is as long as all the other segments taken together (5 in the females and 6 in the males). It has a more or less distinct neck (petiolate) in *Parischnogaster*, *Metischnogaster* and *Anischnogaster* while it is not petiolate in the other genera. This gives these wasps a rather slender silhouette and allows them to touch the tip of the abdomen with their mouth parts when they bend their gaster ventrally. This is particularly important in the egg-laying manoeuvre.

The sixth gastral sternum in the females does not present any cuticular structure at the anterior margin, similar to the "van der Vecht organ", a modified cuticular area with dense hairs that is characteristic of female polistine and vespine wasps (van der Vecht 1968; Spradbery 1975). The males possess an additional segment to the females and in *Parischnogaster*, *Metischnogaster* and *Cochlishnogaster* (Carpenter and Starr 2000) the anterior parts of some gastral tergites (III to VI) are marked with whitish bands which are extremely evident when the abdomen is extended. The tip of the gaster, viewed from the side, is usually pointed in the females and slightly convex, ventrally, in the males (Fig. 2.10), thus distinguishing the two sexes even in those genera (such as *Liostenogaster*) where males and females are very similar in colouration and morphology.

In several genera, the males present more or less modified cuticular structures on their gastral terga: a series of holes associated with cuticular vessel reservoirs (of unicellular glands) on the anterior margin of the 3rd gastral tergum (some *Parischnogaster*, *Stenogaster*, *Eustenogaster*), a large hairy groove on the second (some *Parischnogaster*), transverse bulges ("scrapers") on the anterior margin of 3rd, 4th, 5th and 6th terga (*Liostenogaster*) or peculiar cuticular structures delimiting an internal cavity at the anterior part of the 4th and 5th terga (*Metischnogaster*) (see Fig. 2.21). These structures are connected to tegumental glands.

2.1.3.1 Segments Connected with the Reproductive Apparatus

The last abdominal segments are not visible externally. Their appendages form the stinging apparatus in the females and copulatory apparatus in the males.

Stinging Apparatus

The sting derives from the eighth, ninth and tenth abdominal segments of the female and, when retracted, is contained within a pouch formed by the sixth and seventh segments. Most of the structure is formed by the lateral sclerites of the ninth segment. Accurate descriptions of the sting in *Vespula* are reported by Spradbery (1973) and Edwards (1980) and to these I refer for comparison with the sting of one of the more common species of hover wasps, *Liostenogaster flavolineata*, a sketch of which is given in Fig. 2.11.

To summarise, we can observe that the oblong and quadrate plates (the sclerites to which the muscles moving the sting sheath and lancets are attached) are much

Fig. 2.10 Lateral view of gaster of a female and a male *Parischnogaster*

smaller than those of *Vespula* and other social wasps (cfr Spradbery 1973, plate IIIb): this may account for the reduced efficacy of the sting apparatus in colony defence when compared with Polistinae and Vespinae. In dead hover wasps the extruded sting can sometimes be seen pointing dorsally, while in other wasps it points ventrally (see Fig. 2.12).

Sting extrusion is certainly related to the contraction of the sting muscles and reflects how the use of the sting in these wasps somewhat differs from the others. In fact behavioural observations reveal that hover wasps tend to use their sting by moving it like an externally directed slap delivered with the back of the laterally bent gaster. This contrasts the stinging movements of a polistine or a vespine wasp in which the abdomen is bent ventrally. It seems obvious that the position of the sting and the stinging behaviour in hover wasps are more suitable for slapping small predators (such as ants) away from the nest rather than injecting venom into the tissues of a vertebrate predator.

The sting lancets lack barbs (Fig. 2.13c), which are, on the contrary, found in *Polistes* and *Vespula* albeit much smaller than those of honeybees The stylet is rather squat (Fig. 2.13b). A recent survey on the functionality of the sting in stenogastrine wasps, together with consideration of the enhanced tubular structure of the stylet (see Fig. 2.13a) that limits alternating movements of the lancets, suggests that it is more suitable for convoying outside the Dufour's gland secretion (see later) rather than venom into the skin of Vertebrates (Fortunato and Turillazzi 2012).

At dissection of a sample of 26 females of *Anischnogaster laticeps*, collected from 24 nests, Mike Hansell and I (Turillazzi and Hansell 1991) found two distinct sting length groups. Nineteen females had a sting length ranging from 1.40 mm to 1.60 mm while seven had a sting length ranging from 1.68 mm to 1.80 mm but a body size similar to that of short sting females. Jim Carpenter confirmed to us that, as there was insufficient degree of morphological differences between the two groups of females, this finding should be explained by intra-species polymorphism. The fact that long sting females were not found on initial nests consisting of only one or two cells, led us to offer the hypothesis that this kind of female could be a sort of "social parasite" that usurp established nests (Turillazzi and Hansell 1991).

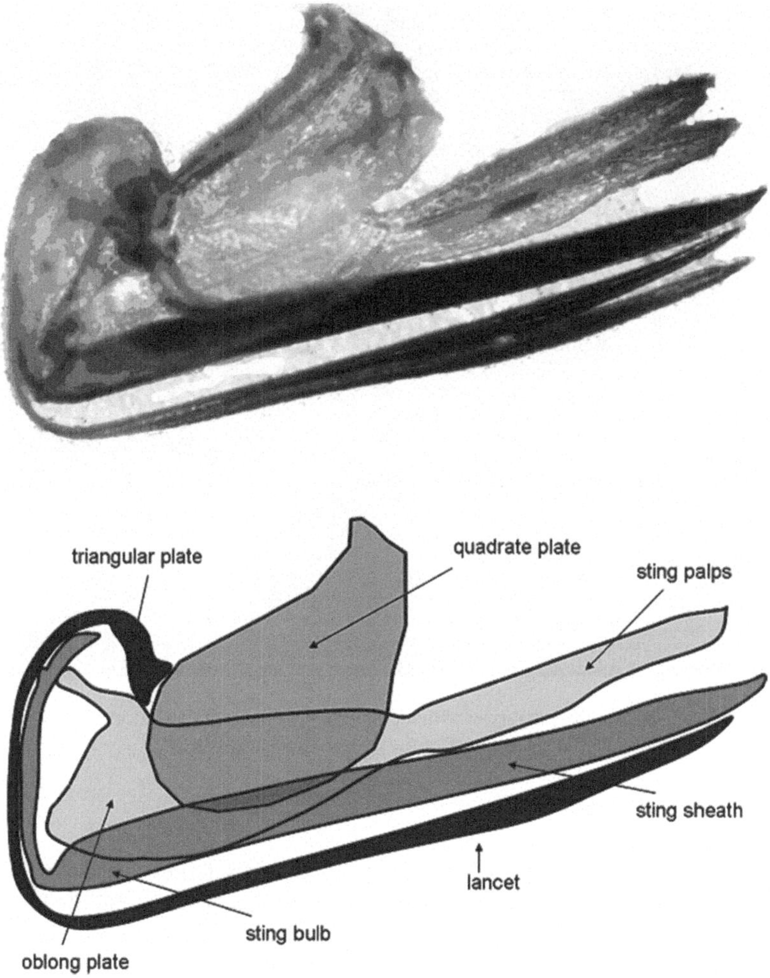

Fig. 2.11 Sting apparatus of a female *Liostenogaster flavolineata*

Fig. 2.12 Extruded sting of a dead female *Liostenogaster flavolineata* (*left*) and of a dead female *Polistes dominulus*

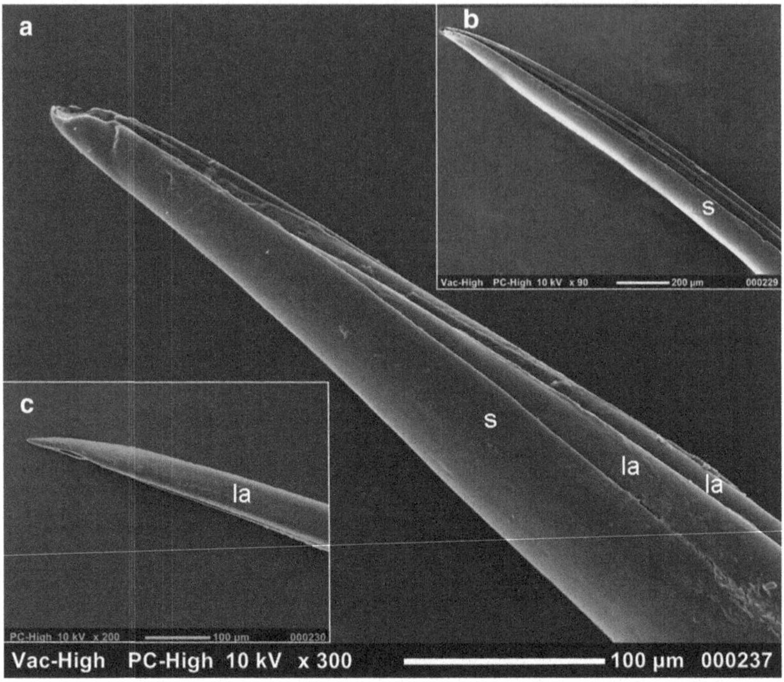

Fig. 2.13 (**a**) tip of the sting, (**b**) stylet without lancets, (**c**) and tip of one lancet of *Liostenogaster flavolineata*. Note the "tubular" structure of the stylet. *s* stylet, *la* lancet

Male Copulatory Apparatus

The male copulatory apparatus consists of two paired sclerites, the gonostipes (or parameres), which enclose an elongate penis (aedeagus) and two mobile lobes, the volsellae, which are divided posteriorly into the digitus and the cuspis (Fig. 2.14). Each gonostipe has long spines in all the genera with the exception of *Parischnogaster* where they are leaf shaped and quite flattened posteriorly with a very small spine (Fig. 2.15).

Males can use the two parameral spines (Fig. 2.16) as a double "pseudo-sting" to deliver painful wounds.[1] The evolutionary meaning of this character waits to be clarified, as many Eumeninae present the same phenomenon (see Spradbery 1973), while other social wasps do not.

The aedeagus is mainly straight but is curved dorsally in *Parischnogaster*.

[1] There is an amusing article by Chris Starr published in Sphecos in 1984 (a newsletter which for several years used to be an excellent means of communication between Hymenopterists) in which the author describes his nasty experience with stings delivered by *Eustenogaster* males in the Philippines.

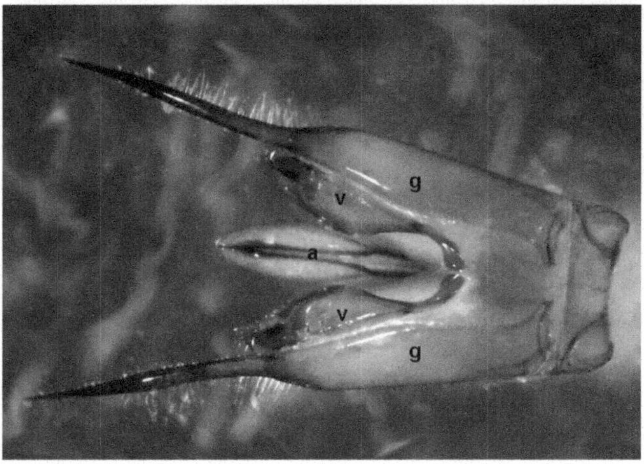

Fig. 2.14 Aedeagus (a) of a male *Liostenogaster flavolineata* and the gonostipes (parameres) (g) and the vosellae (v) in ventral view

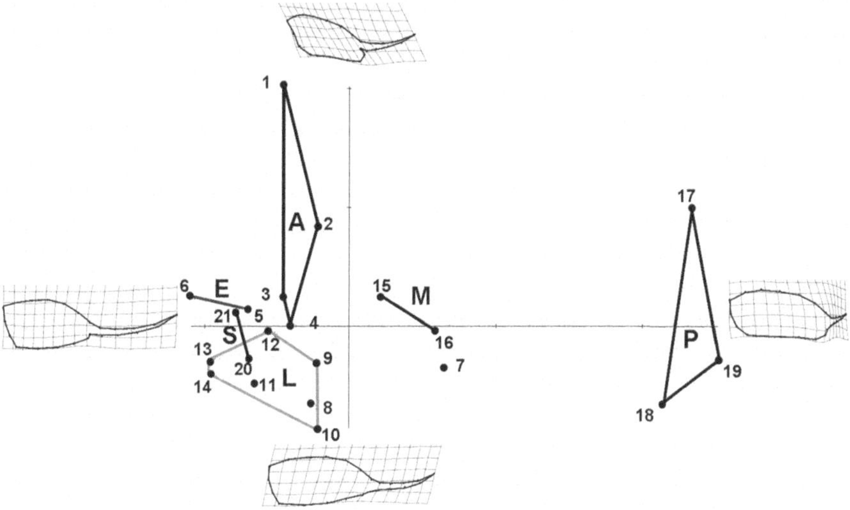

Fig. 2.15 General graph of parameral shape of 22 species belonging to 7 genera obtained with TPS analysis of the landmark points. A = *Anischnogaster* (1 = *A. dubia*, 2 = *A. iridipennis*, 3 = *A.laticeps*, 4 = *A. spilaspis*). E = *Eustenogaster* (5 = *E. luzonensis*, 6 = *E. nigra*). L = *Liostenogaster* (8 = *L. abstrusa*, 9 = *L campanulae*, 10 = *L filicis*, 11 = *L. pardii*, 12 = *L. topographica*, 13 = *L. tutua*, 14 = *L.vechti*). M = *Metischnogaster* (15 = *M cilipennis*, 16 = *M. drewseni*). P = *Parischnogaster* (17 = *P. depressigaster*, 18 = *P. mellyi*, 19 = *P. nigricans*). S = *Stenogaster* (20 = *S. concinna*, 21 = *S macilenta*). 7 = *Holischnogaster* (=*Parischnogaster gracilipes*)

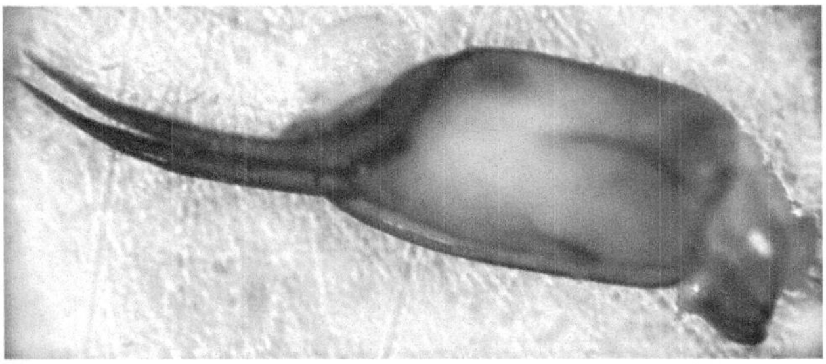

Fig. 2.16 Lateral view of gonostipes of a male *Liostenogaster flavolineata* with the long parameral spines

2.2 Internal Anatomy

The internal anatomy of hover wasps is normally similar to that of other Hymenoptera. This is treated in more general books of Entomology and, in particular for wasps, in the books by Spradbery (1973) and Edwards (1980). At any rate, specific research on the characteristics of the principal systems is needed and at present is limited to the description of exocrine glands.

Modern techniques based on the use of X-rays permit 3D reconstruction of the internal organs of small animals. I used these techniques to visualise and comment on the internal structures of the head, thorax and abdomen of females and males of some species of hover wasps. The figures are self explanatory and evidence the general internal organisation of these wasps while describing the main anatomical systems in the order used by previously cited authors. No differences from other wasps came to light in the circulatory and respiratory systems, which have been little studied in any case (Edwards 1980). Substantial similarities in the nervous system to that of other social wasps are obvious so I redirect the reader to the above-cited reviews. However, in order to furnish more detailed information on the main organs that can be encountered during a dissection aimed at evaluating the social role of an individual, I will dedicate some pages to describing the alimentary system and other noteworthy particularities of these wasps that can be found in the reproductive system and in the exocrine glands complex.

2.2.1 *Alimentary System*

The first part of the alimentary canal, the foregut of ectodermal origin, is divided between head, thorax and gaster. In the head the mouth, delimited by the mandibles, receives saliva from a duct coming from the glands in the thorax. The epipharynx prevents large-sized particles from penetrating further into the intestine. An enlarged pouch behind the pharynx serves to collect various sorts of debris (Fig. 2.17).

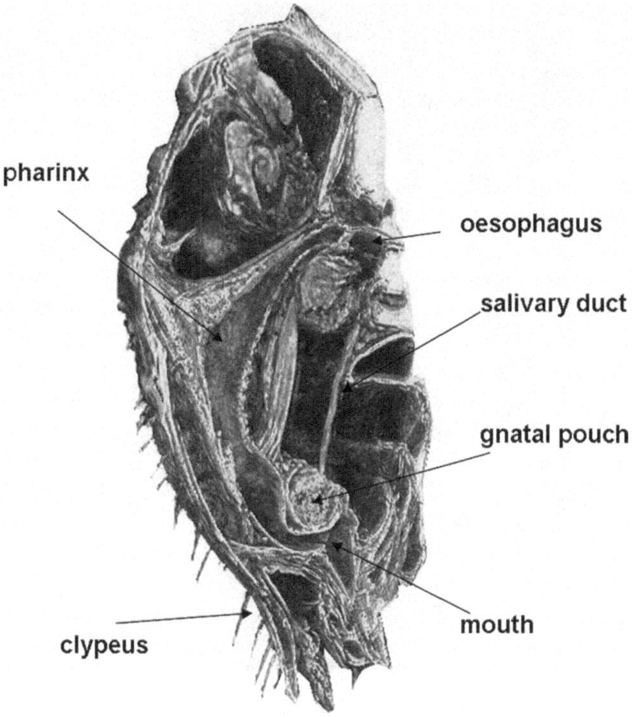

Fig. 2.17 Sagittal section of the head of *L. flavolineata* showing the first part of the alimentary canal (courtesy of Francesca Loglio, CRIST University of Florence)

Before leaving the head capsule, the pharynx becomes the oesophagus, a thin tube which passes through the entire thorax and enters the gaster.

At the level of the gaster, the alimentary canal enlarges into the crop and then into the midgut, which can be recognised by its ringed surface. At the end of the ventriculus are inserted various Malpighian tubules; then the hindgut begins which is formed by an enlarged rectum and the anus. The rectum, which is characterised by longitudinal rectal glands in Polistinae and Vespinae, appears smooth in *Liostenogaster* and *Parischnogaster*.

Comparative studies of the internal anatomy of hover wasps are still lacking but, as far as the alimentary canal is concerned, we can say that it is similar, in its general appearance, to that of other social wasps. The crop in particular, which can be considered a "social stomach" since it allows social Hymenoptera to regurgitate liquid food during their interactions with other members of the colony, is similar in a *Liostenogaster vechti* female, a female of a species of *Polistes* (Fig. 2.18) or a *Vespula* worker. Important differences, however, seem to occur at the hindgut level and wait for an in-depth analysis. In particular the structure of the anus, which looks like a "duck beak" and can be everted in part (unpublished observation), makes me think that it can be used for the deposition of marking pheromones (see also Sect. 3.4.5).

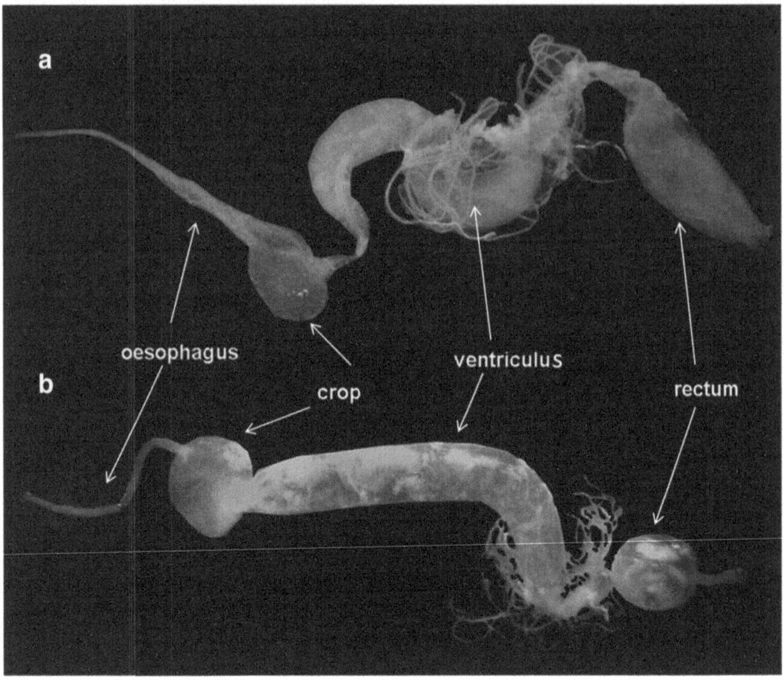

Fig. 2.18 Dissected alimentary canals of (**a**) *Liostenogaster vechti* and (**b**) *Polistes dominula* (courtesy of Iacopo Petrocelli)

2.2.2 Reproductive System

2.2.2.1 Female Reproductive Organs

The female reproductive apparatus is composed of an ovary formed by six polytrophic ovarioles (as in many other wasps) where developing eggs alternate with groups of nutricial cells. A spermatheca, where the sperm of the males is stored after mating, opens into the common oviduct. In several species, the spermatheca, which is important to check to ascertain whether a female has been fertilised or not, is completely enveloped by fat bodies and hardly visible during dissection of the abdomen (Fig. 2.19).

In the aculeate Hymenoptera, the venom apparatus originates from the transformation of the parts which constituted the ovipositor in the primitive Hymenoptera. The duct, originating from the reservoir of the venom glands (two tubular glands which merge together at their entrance into the reservoir) opens into the sting. This is extruded and bent dorsally during egg deposition while the egg passes along the vagina.

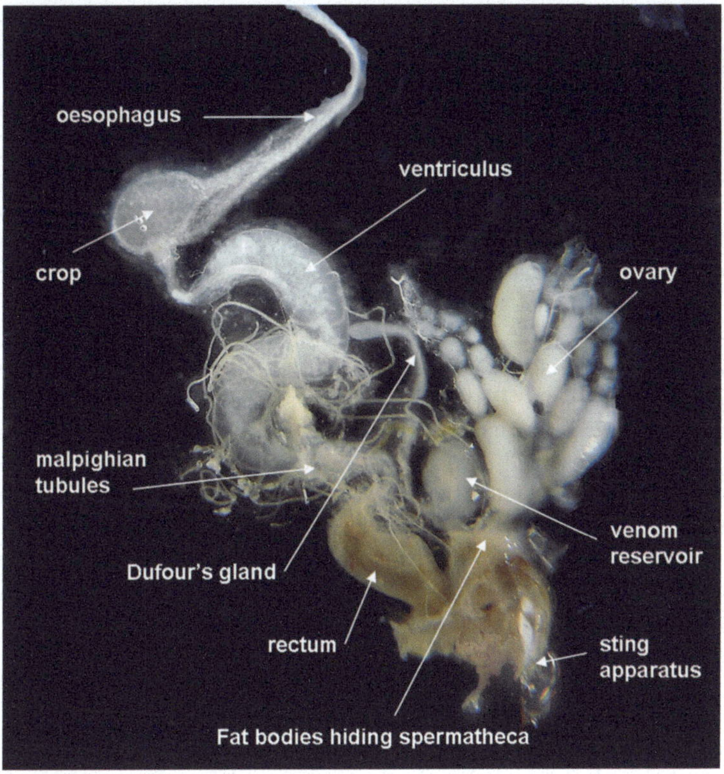

Fig. 2.19 Dissection of the viscera of a female *Liostenogaster flavolineata* showing the principal structures (courtesy of Iacopo Petrocelli)

Exocrine Glands

Various exocrine glands are associated with the reproductive apparatus. In the female these are the Dufour's gland, the venom glands and those associated with the spermatheca. These latter seem to produce a secretion to keep the stored spermatozoa alive.

The Dufour's Gland

The Dufour's gland is tapered and well developed in all the genera. It is much larger than in other wasps and produces a particular, abundant secretion which is quite important in the biology of these wasps (see later) and consists mainly of hydrocarbons (Keegans et al. 1992, 1993). The ultrastructure of this gland has been described by Delfino et al. (1988) in *Parischnogaster mellyi* and *P. alternata*. In these species the gland appears as a tapered tubule which includes two component parts, namely a proximal stalk and a distal bulb. Both tracts consist of a monolayered secretory epithelium and are sheathed by a muscle layer. The stalk possesses a slender lumen and an interrupted contractile sheath, while in the bulb the lumen appears enlarged and the muscle layer becomes continuous.

The gland opens in the ventral wall of the sting bulb into the space comprised by the two lancets, but also in direct connection with the vaginal channel. Fortunato and Turillazzi (2012) recently suggested that the particular position of the gland exit and the tubular morphology of the sting (see Fig. 2.13) allows a small amount of the secretion to be added to the egg during oviposition (which serves to attach the egg to the bottom of the cell), but can be also be channelled to the exterior via the sting so that it can be collected in larger quantities by adult females for larval rearing or constructing ant guards for the nest.

The presence of egg-size drops of a whitish, jelly like substance on the eggs and small larvae of various species had already been reported by the very first students of these wasps: Williams (1919), Jacobson (1935), Pagden (1958), Iwata (1967)) and Spradbery (1975). The first author thought it was of "vegetal origin", the second and third authors presumed it was of "insect origin", while Iwata and Spradbery suggested that it was secreted by the wasps themselves. Hansell (1981) rightly suggested that this was a secretion of the Dufour's gland and this was confirmed by Keegans et al. (1992, 1993) who demonstrated that the content of the Dufour's gland and the patches of substances deposited on the eggs (and also that forming the ant guards in the nests of some species) are substantially the same mixture of hydrocarbons and emulsifying agents.

The following functions for the abdominal secretion of Stenogastrinae on eggs and young larvae have been proposed (1) for attaching the egg to cell walls (Sakagami and Yamane 1990); (2) for nourishing young larvae (Williams (1919), Jacobson (1935), Pagden (1958), Iwata (1967)) and Spradbery (1975), Hansell (1981)), for which there is rather weak evidence, at least in *Parischnogaster* (Turillazzi 1985a); (3) a medium permitting the attachment of the young larva to the bottom of the cell and subsequent rearing (Turillazzi 1985a), and (4) for partially protecting the eggs and young larvae, which has been confirmed in *Parischnogaster* and *Liostenogaster* (Turillazzi 1985a, 1994). The only other context in which abdominal secretion is known to be used is the construction of ant guards (Turillazzi and Pardi 1981; Sledge et al. 2000). Collection processes are reported in Chaps. 3 and 6.

The Venom Gland

The venom gland is formed of two tubules, the ultrastructure of which has been described by Delfino et al. (1997). Histological analysis indicates remarkable regionalisation of the venom gland tubules: whilst the distal tract manufactures the bulk of the secretion, the proximal portion synthesises further products, perhaps single molecules, which are added to the venom before it reaches the muscular storage sac of the sting apparatus. Recently we confirmed (Petrocelli and Turillazzi unpublished) that hover wasps also possess a convoluted gland inside the venom reservoir such as that described in other social wasps, polistine and vespine (*Polistes, Dolichovespula, Vespula, Vespa*) (Schoeters and Billen 1995); this is formed by the

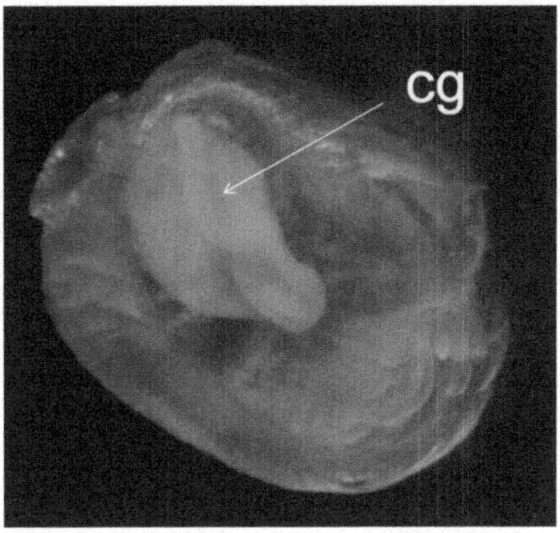

Fig. 2.20 Section of the venom reservoir of *L. vechti* showing the internal convoluted gland (cg) (courtesy of Iacopo Petrocelli)

tubular glands which produce a sort of flattened structure which folds itself inside the lumen of the quite muscular reservoir (Fig. 2.20) (Petrocelli et al. submitted).

The Venom

The venom is a quite complex secretion. While in the solitary species of wasps, such as the Eumeninae, its main function is to cause paralysis of the prey which is then consumed by the developing larvae of the wasps, in social species it is the weapon for killing small enemies and conspecifics and for causing irritation and painful reactions in large vertebrate predators to keep them away from the colonies. However, venom can also contain pheromonal substances which induce and coordinate alarm reactions in a colony while it has recently been demonstrated that in polistine and vespine wasps it may have other important functions, such as furnishing protection against pathogens (Turillazzi et al. 2006). While the composition of the venom of other social wasps is known, at least for the major components, almost nothing is known about the chemical composition of the venom of hover wasps. Dani et al. (1998) used Gas Chromatography coupled with Mass Spectrometry techniques to analyse the volatile compounds of the venom from seven species belonging to three different genera (*Parischnogaster*, *Liostenogaster* and *Eustenogaster*) revealing a mixture of linear alkanes and alkenes with chain length ranging from C11 to C17 in all the species. The composition of the mixture was consistent among conspecifics but clear differences appeared between different species. For example, different spiroacetals (Dani et al. 1998) were found in species belonging to the genus *Parischnogaster*. Some compounds proved to be similar to those found in the venom of other social wasps or even bees, but the biological function of the volatiles found in the venom sac of hover wasps is still unclear (Dani et al. 1998). Preliminary experiments carried out in nature on *Liostenogaster flavolineata* and *P. striatula* did not resolve the issue of whether venom sac volatiles induce alarm or dropping behaviour from the nest, while further experiments led on *P. mellyi* and *P. alternata* kept in captivity seem to exclude the existence of a form of chemical alarm communication in these wasps (Landi et al. 1998) (see also Chap. 5).

Recently Baracchi et al. (2010) analysed the mean molecular weight of the a-polar component (from 900 to 3,000 Da) of the venom of ten species belonging to the three genera mentioned above, demonstrating clear interspecific differences and similarities which can be used as tools for the study of chemosystematics and chemical ecology.

In another recent paper, Baracchi et al. (2012) demonstrated that the venom and methanol cuticular extracts of females of some species of hover wasps have antimicrobial properties. The authors think that, as in other social wasps, this antimicrobial property is due to certain venom peptides which are also spread on the cuticle during self-grooming.

However, nothing is yet known regarding the function of these substances or about the existence of a high molecular weight component that in other social wasps is represented by enzymes, which are also important allergenic agents of venom. The only thing that I can say at present is that while I have quite strong allergic reactions to the stings of polistine and vespine wasps, I have no reaction at all to the sting of hover wasps. This accounts for important differences in a putative high molecular weight component of the venom of the three groups of wasps.

2.2.2.2 Male Reproductive Anatomy

As in other Hymenoptera, the male reproductive apparatus consists of gonads and associated organs.

Exocrine Glands

In male *Parischnogaster* exocrine glands (which probably serve a special function during mating) are associated with the parameres. Males of various species have cuticular glandular apparatuses connected to the gastral terga (Turillazzi and Francescato 1990). In *Parischnogaster striatula*, *P depressigaster* and *P. alternata* this apparatus resides under a hairy groove in the middle of the second gastral tergum (Fig. 2.21). In *Parischnogaster mellyi* and *P. nigricans serrei* males have tegumental glands clustered along the anterior margin of the third gastral tergum (Turillazzi and Calloni 1983; Delfino et al. 1992). In *Liostenogaster* various species present clusters of tegumental glands which open in the intersegmental membrane between the 4–5–6 gastral terga. In *Eustenogaster* there is again a cluster of unicellular glands under the margin of the third gastral tergum. In *Stenogaster* and *Anischnogaster* we found clusters of glands under the third gastral tergum (Francescato et al. 1993). Male *Metischnogaster* instead present peculiar cavities, underlying the whitish tergal bands which are exposed during patrolling, that are delimited ventrally by a layer of secretory cells. At present these tegumental structures wait for further investigations (Turillazzi and Francescato 1990). No information is available for species of *Cochlishnogaster*.

These glands produce secretions that probably play a role in the mating system of the species; secretions are emitted onto the integument through pores and

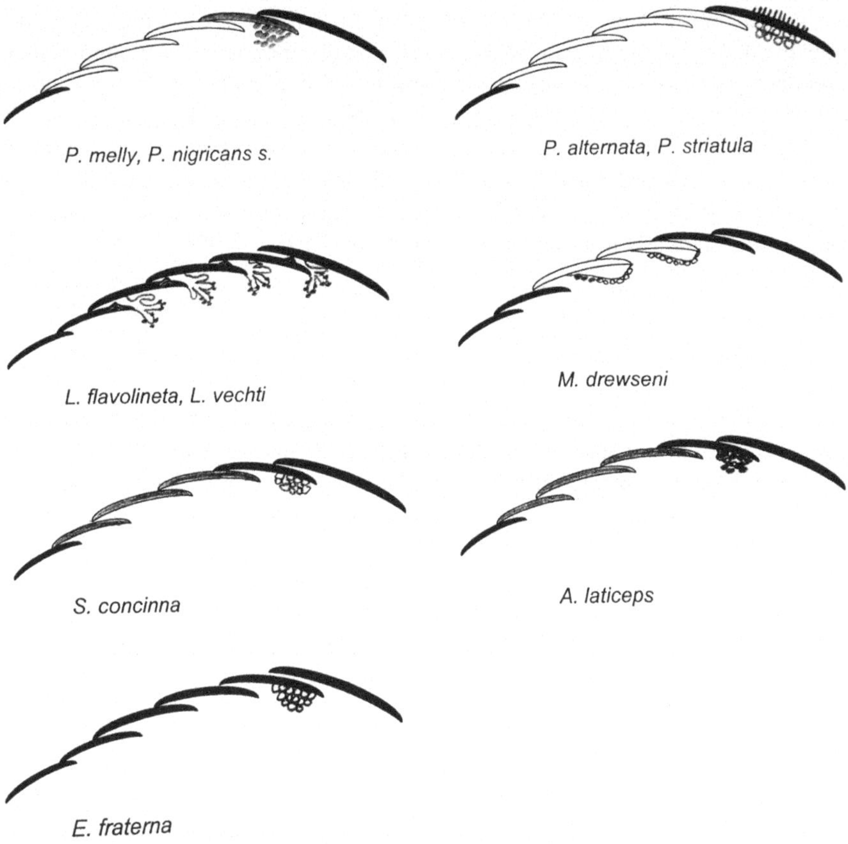

Fig. 2.21 Schematic representation of the disposition of clusters of tegumental glands along the gastral terga of males of various species and genera. Whitish stripes are present on the terga of male *Parischnogaster* and *Metischnogaster*

sometimes directly in connection with tegumental structures which are suitable for the diffusion of the secretion itself.

2.2.3 Other Exocrine Glands

2.2.3.1 Head Glands

A number of exocrine glands occur in the wasp head; we do not know the function of many of them. Males also bear glands, which probably play an important role during mating, on their antennae. At present, studies on head glands in hover wasps are non-existent with the exception of preliminary studies by Delfino et al. (1998)

performed on ectal mandibular glands of *Liostenogaster flavolineata*. The mandibular glands are the largest glands in the head. In *Polistes* females the secretion is probably defensive but in the males of *Polistes major* it seems they are important in reproductive behaviour (Wenzel 1987). Similarly in the males of *Parischnogaster* and *Metischnogaster* they are probably involved in the mating system of the species. In *Parischnogaster mellyi* males Beani et al. (2002) found ectal mandibular glands three times wider and longer than in the females, although mandible size is not sexually dimorphic. The mandibular glands of male *Parischnogaster jacobsoni* can be seen in the 3D head reconstruction in Fig. 2.22.

2.2.3.2 Thoracic Glands

The most important exocrine glands in the thorax are the salivary glands. These are formed by clusters of unicellular glands (Fig. 2.23) that emit their secretion into ducts which merge together to form a single duct opening into the salivarium, the last part of the mouth.

2.3 Morphology and Anatomy of the Immature Stages

2.3.1 The Egg

The egg of all species is usually shaped like a rough sausage and white to yellowish in colour. Unlike other social wasps, which attach one of the egg extremities to the cell walls, hover wasps usually attach their eggs by their convex part (Fig. 2.24). The attaching medium, at least in *Parischnogaster*, is a drop of secretion the composition of which is quite similar to that produced by the Dufour's gland (see Sect. 3.5.2). The sizes of eggs for various species are reported by Iwata (1967) and Kojima (1990) and range in length from 0.92 mm (on average) in *Parischnogaster nigricans serrei* to a maximum of 1.8 mm in *E. eximia* (Krombein 1991) (see also Table 1 in Kojima 1990). At present no study exists on the ultrastructure of the egg shell and on its possible systematic importance.

The eggs of most species are usually covered with a gelatinous substance which is mainly produced by the Dufour's gland (Keegans et al. 1992, 1993). As we have seen, this secretion serves various functions in the economy of a hover wasp colony and is composed mainly of a mixture of hydrocarbons.

Oviposition behaviour is described in Chap. 3 (Turillazzi 1985b).

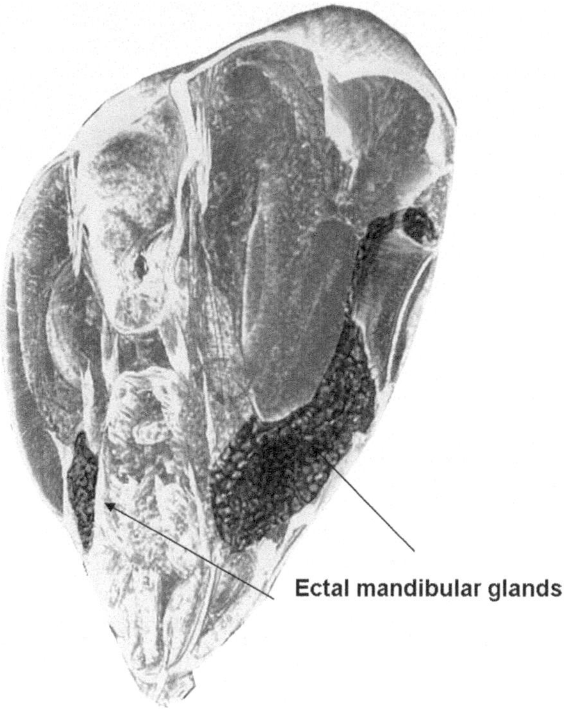

Fig. 2.22 3D Reconstruction of the head of a male *Parischnogaster jacobsoni* seen from the back and cut on transversal plane. The extension of the ectal mandibular glands is *highlighted* (courtesy of Francesca Loglio, CRIST University of Florence)

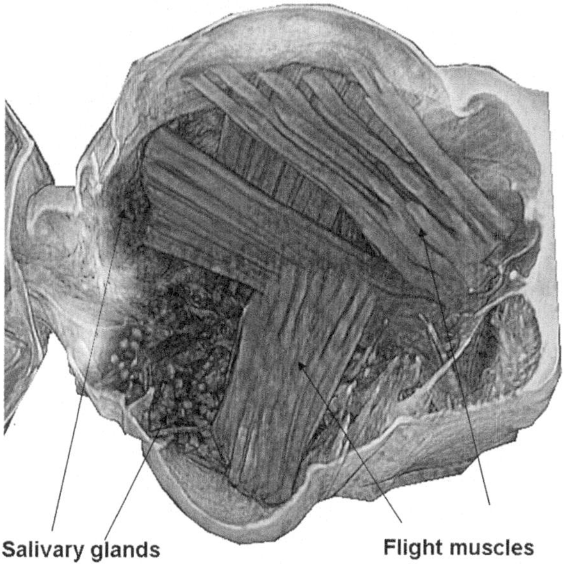

Fig. 2.23 Sagittal section of the thorax of a female *L. flavolineata* showing the flight muscles and the clusters of salivary glands

Fig. 2.24 Eggs of *Liostenogaster flavolineata* with and without Dufour's gland secretion

2.3.2 The Larva

The larva hatches after a varying number of days depending on the species. In *Parischnogaster* hatching occurs, on average, 7 days after egg deposition. The larval development consists of four stages in species of *Parischnogaster*, *Eustenogaster*, *Liostenogaster* and *Anischnogaster* (Hansell 1982, 1986; Samuel 1987; Turillazzi 1985c, 1990; Turillazzi and Hansell 1991). Iwata (1967) on the other hand reported instead five larval instars in the Stenogastrinae (Fig. 2.25).

Kojima (1990) reported the distribution of head width of larvae for four species (*L. vechti*, *E. calyptodoma*, *A. iridipennis* and *P. mellyi*); only four groups were found for the fourth species while for the second ($N = 15$) and third species ($N = 10$) there were insufficient samples to deduce the presence of four or five instars, despite the claims of the author to have identified a fifth, initial, larval instar. In *L. vechti* the sample is more numerous ($N = 94$) and indicates the presence of five instars. I looked at the same species, measuring the head width of 110 larvae, including some still in their egg shells, and the distribution I found showed four main clusters (none of which statistically different from a normal distribution) (Turillazzi 1990). Samuel examined a very large sample in *L. flavolineata* ($N = 190$) and similarly concludes that the distribution "strongly indicates that there are four larval instars during the course of larval development" in this species (Samuel 1987). Following Cooper (1966), Carpenter (1988) warns that the definitive solution to this dispute can only be reached through direct observation of the first instar by witnessing hatching and an almost continuous observation until the emergence of the imago and the collection of all the exuviae. Thus the uncertainty on the matter remains even if the concluding presence of only four larval instars, a character which would differentiate the Stenogastrinae from all the other vespid wasps, would not in any case be important for clarifying their phylogenetic relationships.

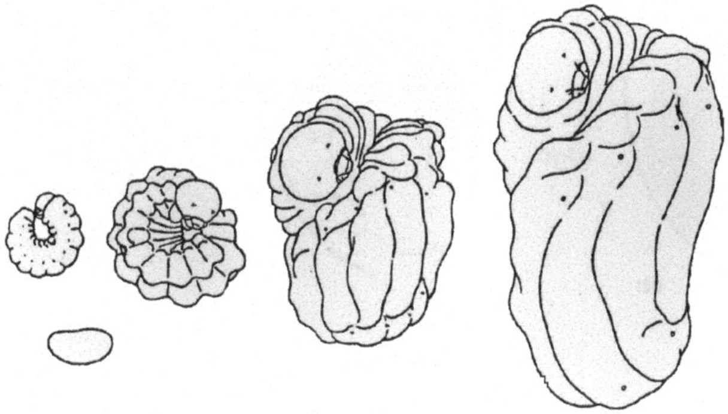

Fig. 2.25 Larval stages of *Parischnogaster* sp. (from Iwata 1967 redrawn)

In *Parischnogaster*, immediately after hatching the larva curls itself around the mass of gelatinous secretion, produced by the Dufour's gland of the parent, which covers the eggs (Fig. 2.26).

During the very first days the larvae feed mainly on liquid food and on tiny cuds of regurgitated chewed food which the adults place on the mass of secretion. This is not in itself a food but rather a substrate which permits the larvae to remain attached to the cell as well as a "dish" to collect the supplements of the adults. In some species pieces of malaxated food particles are often visible mixed with the secretion placed on still un-hatched eggs or quite small larvae.

The more the larva increases in size, the less secretion is added by the adults, so that a last instar larva remains curled in the cell pushing its back against the cell walls (Fig. 2.27a) (Turillazzi 1985a).

This is another important difference with respect to the other social wasps, where the larvae keep their body distended along the longitudinal axis of the cells and present only their heads at the cell opening. When the adult hover wasps nourish their larvae, they touch the internal sides of the larva with their mandibles; the larva then opens itself up like a kind of sphincter to receive the cud of food (see Fig. 3.9). In the other social wasps, on the contrary, the adults supply the food directly into the mouth of the larva. In a recent experiment I observed that a last instar larva of *Parischnogaster mellyi* was capable of feeding itself autonomously on freshly killed small insects (such as fruit flies or mosquitoes) inserted with thin forceps within its coil (Fig. 2.27b).

Kojima (1990) lists a number of general characters of external morphology of the last instar larvae of hover wasps: a fusiform body, a clypeus much wider than tall, spinneret without raised lip, the presence of 10 circular spiracles and of pleural lobes, and the abdominal segments from 1 to 6 with dorsal lobes. The head capsule is un-pigmented with the exception of *L. flavolineata* (cf. Carpenter 1988). In all the species Kojima examined, the larval antennae had a papilla shaped differently from Polistinae and Vespinae. The larval mandibles are tridentate and are quite large with respect to those of other social wasps. The first larval spiracle is distinctly larger than the others (with the exception of *Eustenogaster*) and this is also considered an

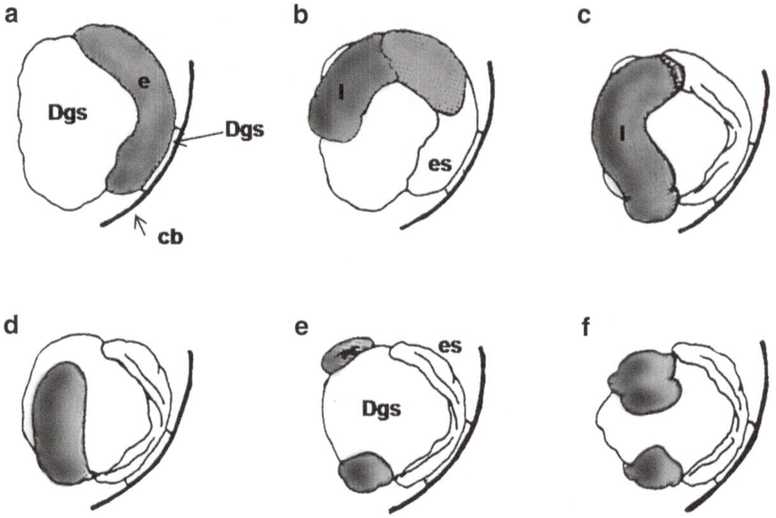

Fig. 2.26 Hatching sequence of a larva of *Parischnogaster mellyi* (from Turillazzi 1985a, redrawn) (*e* egg, *es* egg shell, *cb* cell bottom, *l* larva, *Dgs* Dufour's gland secretion)

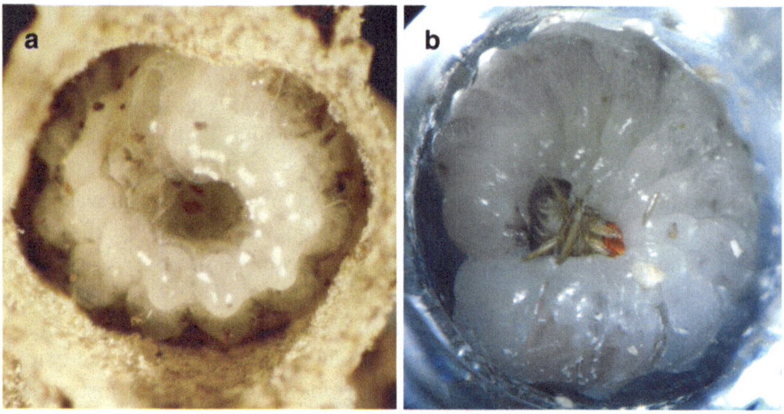

Fig. 2.27 a) A well-developed larva of *Liostenogaster flavolineata* curled in its cell. b) Larva of *Parischnogaster mellyi* eating an entire fruit fly experimentally inserted into its coil.

autapomorphy of the group. Another autapomorphy of the subfamily are the long setae found on the dorsal surface of the first thoracic segment (Kojima 1990).

According to Spradbery (1975), stenogastrine larvae have well-developed salivary (labial) glands comparable to those of vespine larvae, but he does not report any measurements or images. Spradbery observes that in other social wasps the glands have a social function to secrete sugars, amino acids and proteases. The adults suck these up during trophallactic exchanges and the larvae emit them as drops of fluid after antennal and mandibular stimulation by the adults. As he failed to obtain regurgitation of a drop of fluid after stimulating a larva of *S. concinna*, he

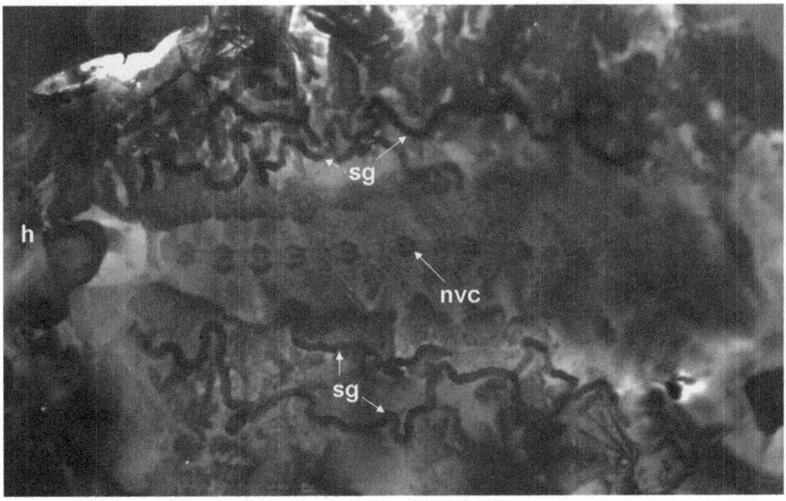

Fig. 2.28 Parts of the salivary glands of a dissected mature larva of *L. vechti*. *sg* salivary glands, *nvc* nervous ventral chain, *h* head. Foregut, midgut and larval malpighian tubules have been removed to permit a better view of the glands (courtesy of Iacopo Petrocelli)

suggests that in the hover wasps these glands function primarily as an organ of digestion and, rarely, for spinning a pupal cocoon.

I have observed and video-recorded various episodes of larval nutrition in different species and especially in *P. mellyi*. In some cases I recorded females without evident food boluses visiting recently fed larvae and inserting their mouthparts within the coiled larvae. The larvae reacted by opening themselves up and liquid appeared within the coils while the wasps seemed to lick up the fluid and clean their mouth parts after contact. In these cases I could not exclude the possibility that the wasp supplied liquid food to the larvae, so decisive evidence of larval–adult trophallaxis is still lacking. However, after observing the different development of these glands in a stenogastrine (*L. flavolineata*) and a polistine wasp (*P. dominula*), I have some doubts even about the possibility of an indirect passage of fluids from larvae to adults in this species. Moreover, dissections (Fig. 2.28) and X-rays surveys of larvae of *P. jacobsoni* and of *L. flavolineata* demonstrated that the glands are not comparable in size with those of a larva of *Polistes* or *Vespula* (see for example Fig. 32 in Spradbery 1973) where larval–adult trophallaxis is one of the main features of social organisation. In any case, after a while the adults can take the food they first left inside the coiled larvae away again; then they can chew it further and redistribute to other larvae (Turillazzi 1985c, personal observation).

2.3.3 The Pupa

When pupating, the larva spins an incomplete cocoon (but some species, such as *S. concinna*, do not spin a cocoon at all—Spradbery 1975) inside the cell, but the cell itself is closed by the adults with the same kind of material used for the construction of the nest. In some species, however, as in all the species included in the genera *Eustenogaster* and *Stenogaster,* the cell opening is only narrowed and not completely closed (Fig. 2.29).Various authors noted that in many genera pupal cells or cells from which adults had emerged were coated with a glistening material. In *E. eximia* Krombein (1991) observes that a delicate glistening film which is applied (as a painting) to the entire interior walls of the pupal cell is actually underlined by strands of fine silk. The author suggests that this could be important for the protection of the developing pupa from the infestation of fungi, the hyphae of which are commonly present in nest material

The operculum is then re-opened some time (2–3 days) afterwards because the adults have to eliminate the larval faeces which have been emitted by the larvae after pupation (Turillazzi and Pardi 1982). This phenomenon is also widespread and analogous in the species of the polistine tribe Ropalidiini where the larval faeces are eliminated through a hole that the adults tear in the back of the cell; in the other social wasps larval stool is simply compressed on the bottom of the cell where it remains, even if the cell is re-used. In the Stenogastrinae, once the larval meconium (Fig 2.30) is eliminated, the hole in the operculum is closed again by the adults. This can be easily seen by colouring the opercula of recently closed cells and checking the colour patches after some days (this can also be observed in some natural nests such as that presented in Fig. 6.31).

The pupa (and the prepupa) assumes a position in the cell with the body lying along the longitudinal axis and head towards the operculum with the abdomen bent towards the mouthparts (Fig. 2.31). The pupal metasoma bent at the junction between the first and second segment is a general feature of hover wasps and it also occurs in some eumenines. The character is differently evaluated by van der Vecht, who considers it as evidence of the phylogenetic proximity between hover wasps and the Eumeninae, and Carpenter (1982) who interprets this as a convergent adaptation related to the long metasomal petiole in the two groups.

Another peculiar character of hover wasp pupae is the presence of a pair of prongs on the mesoscutum. Kojima (1990) concludes that no larval or pupal characters are informative for reconstructing the phylogenetic relationships among genera.

2.3.4 Duration of Immature Development

The development from egg to emergence varies widely depending on the species and environmental situations but information about this is quite limited. Complete

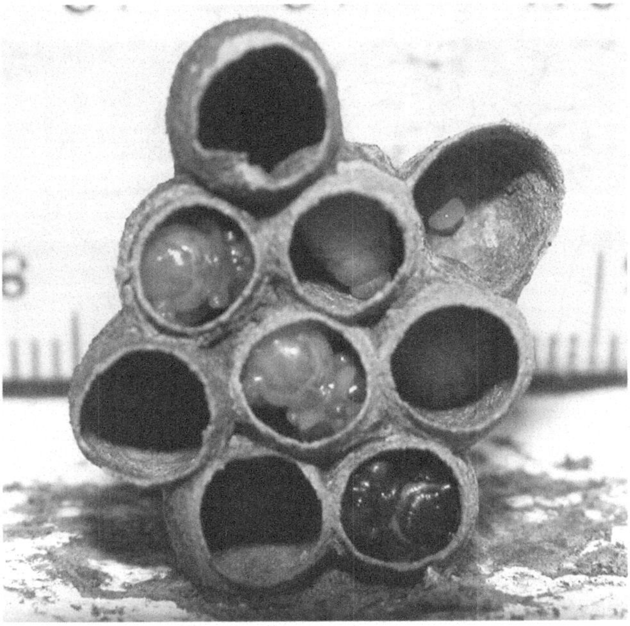

Fig. 2.29 Pupae of *Eustenogaster calyptodoma*: the cell openings are not operculate but only narrowed by the adults

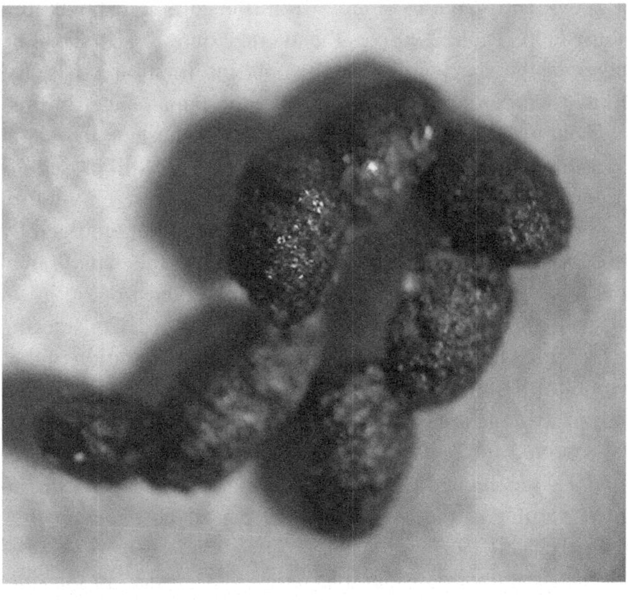

Fig. 2.30 Larval meconium discarded from a nest of *L. flavolineata*

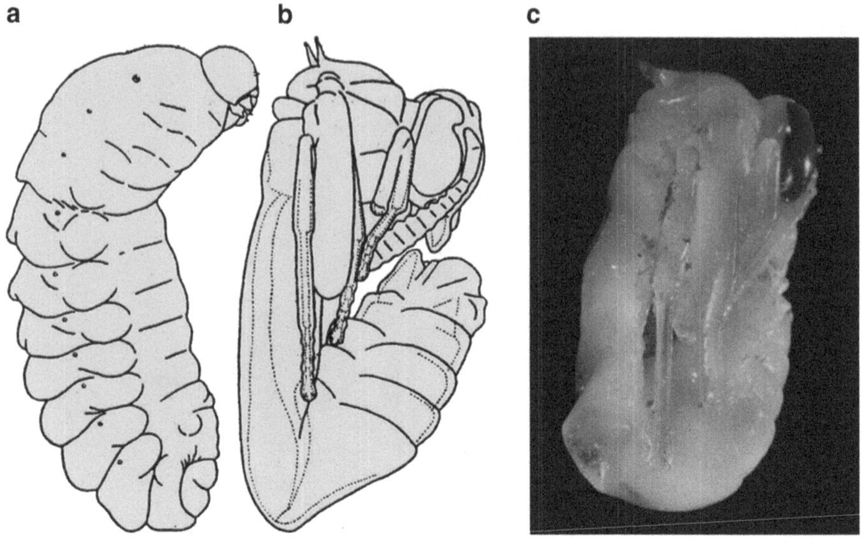

Fig. 2.31 (**a**) Prepupa and (**b**) pupa of *Parischnogaster* sp. (re-drawn from Iwata 1967). (**c**) Pupa of *Parischnogaster* sp.

immature development (from egg deposition to emergence) lasts on average from a minimum of 44.5 days in *P. nigricans serrei* (Turillazzi 1985d) to a maximum of 103 days in *L. flavolineata* (Samuel 1987). *Eustenogaster calyptodoma* presents an average immature development of about 63 days (Hansell 1987). The completely formed adult is evidently able to emerge from the cell, after breaking the operculum, without any help. Table 2.2 gives the average duration (in days) of egg, larval and pupal stages of the species for which some information exists. In the case of *L. vechti* we have scarce data referring only to the pupal stage. In this species we observed that 7 out of 22 pupal cells found in various nests at the beginning of the observation period during our last visit to Malaysia were still closed after 36 days. Four out of 17 cells of which we witnessed the closure were only left by the adults after several days and the pupae of a further 12 had not yet emerged by the end of the observation period. We were only able to establish the actual pupal period for one pupa, which was 37 days. Thus, we know that at least one-third of the larvae observed had a pupal period that lasted more than 36 days. Considering that in the other species for which there is complete information the duration of the pupal stage lasts roughly one-third of the entire immature development, for *L. vechti* we should expect to find an average total developmental period of over 100 days, roughly similar to that reported for *L. flavolineata*..

Krombein (1991) estimated a pupal period of about 21 days in *Eustenogaster eximia* which is similar to the average pupal period observed in the congeneric *E. calyptodoma* (Hansell 1987).

As is obvious, there is a considerable range of variation in the duration of the different phases in all the species (note especially the very long interval in the larval

Table 2.2 Duration in days of the various phases of immature development in six species of hover wasps belonging to four genera. For *L. vechti* the total immature development is inferred (unpublished observations) as three times the length of the pupal phase (considering that this is roughly the ratio found in the other species)

	P. mellyi (Hansell 1982)	*P. nigricans serrei* (Turillazzi 1985d)	*E. calyptodoma* (Hansell 1987)	*L. flavolineata* (Samuel 1987)	*L. vechti* (Turillazzi et al. unpublished)
Egg phase	8.05 (7–10; $N = 5$)	11.39 (3–22; $N = 82$)	42.07	70.02 (25–120; $N = 97$)	
Larval phase	25.71 (11–40; $N = 39$)	16.62 (6–29; $N = 80$)			
Pupal phase	17.08 (14–21; $N = 42$)	17.12 (7–29; $N = 80$)	21.04 ($N = 15$; SD = 2.54)	32.04 (14–47; $N = 85$)	>36
Total immature development	52.00 (32–61)	44.46 (34–66; $N = 75$)	64.00	103.06	>100 (supposed)
Average pupal/total immature development	0.34	0.38	0.33	0.32	

period of *L. flavolineata*), but we must consider that various factors can heavily influence the larval (and pupal) development, starting with food availability, number of nurses per larva, atmospheric conditions etc.. The total immature development in all the species in general lasts longer than in the other social wasps (for example in *V. vulgaris* it averages slightly more and in *V. crabro* slightly less than 35 days—Spradbery 1973) but the very long period (103 days) recorded for *L. flavolineata* is really striking, especially considering that the maximum age reached by a female directly observed in this species was 238 days (Samuel 1987). The length of immature development can have important consequences for the social organisation of the species and could have been a factor favouring the onset of sociality in the group as we shall see later.

References

Baracchi D, Dapporto L, Teseo S, Hashim R, Turillazzi S (2010) Medium molecular weight polar substances of the cuticle as tools in the study of the taxonomy, systematics and chemical ecology of tropical hover wasps (Hymenoptera: Stenogastrinae). J Zool Syst Evol Res 48:109–114

Baracchi D, Mazza G, Turillazzi S (2012) From individual to collective immunity: the role of the venom as antimicrobial agent in the Stenogastrinae wasp societies. J Insect Phys 58:188–193

Beani L, Sledge MF, Maiani S, Boscaro F, Landi M, Fortunato A, Turillazzi S (2002) Behavioural and chemical analysis of scent-marking in the lek system of a hover-wasp (Vespidae, Stenogastrinae). Insect Soc 49:275–281

Carpenter JM (1982) The phylogenetic relationships and natural classification of the Vespoidea (Hymenoptera). Syst Entomol 7:11–38

Carpenter JM (1988) The phylogenetic system of the Stenogastrinae (Hymenoptera: Vespidae). J N Y Entomol Soc 96:140–75

Carpenter JM (2001) New generic synonymy in Stenogastrinae (Insecta: Hymenoptera; Vespidae). Nat Hist Bull Ibaraki Univ 5:27–30

Carpenter JM, Starr CK (2000) A new genus of hover wasps from Southeast Asia (Hymenoptera: Vespidae; Stenogastrinae). Am Mus Novit 3291:1–12

Cooper KW (1966) Ruptor ovi, the number of moults in development, and method of exit from masoned nest. Biology of eumenine wasps, VII. Psyche 73:238–250

Dani RF, Morgan ED, Jones GR, Turillazzi S, Cervo R, Francke W (1998) Species-specific volatile substances in the venom sac of hover wasps. J Chem Ecol 24:1091–1104

Delfino G, Turillazzi S, Calloni C (1988) A light and electron microscopy study on the Dufour gland in *Parischnogaster* (Hymenoptera Stenogastrinae). Zeit Microsc Anat Forsch 102:627–644

Delfino G, Calloni C, Turillazzi S (1992) Ultrastructure of the tegumental glands of the third gastral tergite in male *Parischnogaster mellyi* (Saussure) (Hymenoptera Stenogastrinae). Ethol Ecol Evol 2:55–59

Delfino G, Calloni C, Turillazzi S (1997) Occurrence of class II structural patterns in the venom glands of stenogastrine wasps. J Nat Toxins 6:51–68

Delfino G, Fortunato A, Formigli F, Turillazzi S (1998) Structural and ultrastructural features of the ectal mandibular gland in Liostenogaster (Vespidae, Stenogastrinae). Ins Soc Life 2:45–52

Edwards R (1980) Social wasps, their biology and control. Rentokil Limited, East Grinstead

Fortunato A, Turillazzi S (2012) Dufour's gland and massive use of its secretion drove the evolution of sting's morphology and function in hover wasps (Hymenoptera Stenogastrinae). Arthropod Struct Dev 41:259-264

Francescato E, Turillazzi S, Hansell MH (1993) Glandular apparatus associated with the gastral tergal tegument in males of *Stenogaster concinna* and *Anischnogaster laticeps* (Hymenoptera Stenogastrinae). Insect Soc 40:313–317

Hansell MH (1981) Nest construction in the subsocial wasp *Parischnogaster mellyi* (Saussure) (Stenogastrinae Hymenoptera). Insectes Soc 28:208–216

Hansell MH (1982) Brood development in the subsocial wasp *Parischnogaster mellyi* (Saussure) (Stenogastrinae Hymenoptera). Insectes Soc 29:3–14

Hansell MH (1986) Colony biology of the stenogastrine wasp *Holischnogaster gracilipes* (Van der Vecht) on Mount Kinabalu. Entomol Mon Mag 122:31–6

Hansell MH (1987) Elements of eusociality in colonies of *Eustenogaster calyptodoma* (Sakagami & Yoshikawa) (Stenogastrinae, Vespidae). Anim Behav 35:131–141

Iwata K (1967) Report of the fundamental research on the biological control of insect pests in Thailand. II. The report on the bionomics of Aculeate wasps bionomics of subsocial wasps of Stenogastrinae (Hymenoptera, Vespidae). Nat Life SEA 5:259–293

Jacobson E (1935) Aantekeningen over Stenogastrime (Hym. Vespid.). Ent Med Soc 14:309–318

Keegans SJ, Morgan ED, Jackson BD, Berresford G, Turillazzi S, Billen J (1992) The chemical nature and origin of the abdominal substance of the social wasp *Liostenogaster flavolineata* (Vespidae, Stenogastrinae). In: Billen J (ed) Biology and evolution of social insects. Leuven University Press, Leuven Belgium

Keegans SJ, Morgan DE, Turillazzi S, Jackson BD, Billen J (1993) The Dufour gland and the secretion placed on eggs of two species of social wasps. J Chem Ecol 19:279–290

Kojima J (1990) Immatures of hover wasps (Hymenoptera, Vespidae, Stenogastrinae). Jpn J Entomol 58:506–522

Krombein KV (1991) Biosystematic studies of Ceylonese wasps. XIX: natural history notes in several families (Hymenoptera, Eumenidae, Vespidae, Pompilidae, Crabronidae). Smithsonian Contrib Zool 515:1–41

Landi M, Coster-Longman C, Turillazzi S (1998) Alarm response in a laboratory population of *Parischnogaster mel*lyi (Stenogastrinae). Ins Soc Life 2:59–64

Pagden HT (1958) Some Malayan social wasps. Malayan Nat J 12:131–148

Rohlf FJ (2006) Tps Dig version 2.10. Department of Ecology and Evolution, State University of New York at Stony Brook, New York

Rohlf FJ (2007) Tps Relw version 1.45. Department of Ecology and Evolution, State University of New York at Stony Brook, New York

Sakagami SF, Yamane S (1990) A behavior inventory of the females of two stenogastrine wasps *Parischnogaster mellyi* and *Liostenogaster vechti* (Hymenoptera, Vespidae). In: Sakagami SF, Ohgushi R, Roubik DW (eds) Natural history of social wasps and bees in equatorial Sumatra. Hokkaido University Press, Sapporo

Samuel C (1987) Factors affecting colony size in the stenogastrine wasp *Liostenogaster flavolineata*. PhD thesis, University of Malaya, Kuala Lumpur

Schoeters E, Billen J (1995) Morphology and ultrastructure of a secretory region enclosed by the venom reservoir in social wasps (Insects, Hymenoptera). Zoomorphology 115:63–71

Sledge MF, Fortunato A, Turillazzi S, Francescato E, Hashim R, Moneti G, Jones GR (2000) Use of Dufour's gland secretion in nest defence and brood nutrition by hover wasps (Hymenoptera Stenogastrinae). J Insect Phys 46:753–761

Spradbery JP (1973) Wasps: an account of the biology and natural history of solitary and social wasps. Sidgwick and Jackson/University of Washington Press, London/Seattle

Spradbery JP (1975) The biology of *Stenogaster concinna* Van der Vecht with comments on the phylogeny of Stenogastrinae (Hymenoptera: Vespidae). J Aust Entomol Soc 14:309–18

Starr CK (1984) Rambling remarks on Philippine wasp-mimics. Sphecos 9:11–13

Turillazzi S (1985a) Function and characteristics of the abdominal substance secreted by wasps of the genus *Parischnogaster*. Monit Zool Ital 19:91–99

Turillazzi S (1985b) Egg deposition in the genus *Parischnogaster* (Hymenoptera Stenogastrinae). J Kansas Ent Soc 58:749–752

Turillazzi S (1985c) Brood rearing behaviour and larval development in *Parischnogaster nigricans serrei*. Insect Soc 32:117–127

Turillazzi S (1985d) Colonial cycle of *Parischnogaster nigricans serrei* (Du Buysson) in West Java (Hym. Stenogastrinae). Insect Soc 32:43–60

Turillazzi S (1990) Social biology of *Liostenogaster vechti* Turillazzi 1988 (Hymenoptera Stenogastrinae). Trop Zool 3:68–87

Turillazzi S (1994) Protection of brood by means of Dufour's gland secretion in two species of stenogastrine wasps. Ethol Ecol Evol (special Issue 3):37–41

Turillazzi S, Calloni C (1983) Tegumental glands in the third gastral tergites of male *P.nigricans* and *P. mellyi*. Insect Soc 30:455–460

Turillazzi S, Francescato E (1990) Patrolling behaviour and related secretory structures in the males of some stenogastrine wasps (Hymenoptera Vespidae). Insect Soc 37:146–157

Turillazzi S, Hansell MH (1991) Biology and social behaviour of three species of *Anischnogaster* (Vespidae, Stenogastrinae) in Papua New Guinea. Insect Soc 38:423–437

Turillazzi S, Pardi L (1981) Ant guard in the nest of *Parischnogaster nigricans serrei* (du Buysson). Monit Zool Ital 15:1–7

Turillazzi S, Pardi L (1982) Social behavior of *Parischnogaster nigricans serrei* (Du Buysson) (Hymenoptera Stenogastrinae) in Java. Ann Entomol Soc Am 75:657–664

Turillazzi S, Mastrobuoni G, Dani FR, Moneti G, Pieraccini G, La Marca G, Bartolucci G, Perito B, Lambardi D, Cavallini V, Dapporto L (2006) Dominulin A and B: two new antibacterial Peptides identified on the cuticle and in the venom of the social paper wasp *Polistes dominulus* using MALDI-TOF, MALDI-TOF/ TOF, and ESI-Ion Trap. J Am Soc Mass Spectrom 17:376–383

van der Vecht J (1968) The terminal gastral sternite of female and worker social wasps (Hymenoptera, Vespidae). Proc K Ned Akad Wet Ser C Biol Med Sci 71:411–422

van der Vecht J (1972) A review of the new genus *Anischnogaster* in the Papua region (Hymenoptera, Vespidae). Zool Meded Leiden 47:240–256

Wenzel JW (1987) Male reproductive behavior and mandibular glands in *Polistes major* (Hymenoptera Vespidae). Insectes soc 34:44–57

Williams FX (1919) Philippine wasp studies, part 2: descriptions of new species and life history studies. Bull Exp Sta Hawaii Sugar Planters' Assoc Entomol 14:19–186

Chapter 3
Behaviour

3.1 Female and Male Behavioural Patterns

As we have seen from the description of the general morphology and anatomy of the adults, differences between the sexes are sometimes very slight in most species of hover wasps and fundamentally related to sexual selection factors. In general, males can hardly be distinguished at first sight when they are on their nests and, often, I have been surprised to find I collected colonies composed of individuals of both sexes when I imagined I was going to capture members of one sex only. This usually happens in colonies of species of *Liostenogaster* where the male colouration seems to perfectly mimic that of the female. But, notwithstanding these similarities in external morphology, the two sexes diverge greatly in behaviour. Behavioural patterns have often been utilised as important characters in taxonomy even if they are more difficult to treat than morphological ones and are not always available for all species. Behavioural patterns are usually sequences of singular modules, results of complex biomolecular reactions, phenotypic expressions of the interactions between genome and environment. However, they can be suitable for cladistic analyses if they present definite states and when compared at the same "magnification" in different taxa. One very fitting example is nest construction behaviour, where even the very final output of the behavioural sequence can be compared: i.e. the nest itself. But of this particular behaviour I shall speak in Chap. 6. Here I wish to furnish a survey of the behavioural patterns we have found in hover wasps and define the principal characteristic behavioural sequences which are important in the biology of these insects. To illustrate the various patterns, I shall also refer to videoclips, which are published on line, that can help the reader to compare a given behaviour with the same behaviour observed in other wasps.

Previous authors have already described some of the behavioural patterns of various species of hover wasps; in organising this review I found particularly useful

The video clips referred to in this chapter can be found under http://extras.springer.com or in the supplementary material (ESM) of the electronic version of this chapter.

S. Turillazzi, *The Biology of Hover Wasps*, DOI 10.1007/978-3-642-32680-6_3,

to use, as a referring point, the paper by Soichi F. Sakagami and Soichi Yamane (1990), who traced the behaviour inventory of *Parischnogaster mellyi* and *Liostenogaster vechti*. Of course I shall also use the observations of other authors as well as my own personal observations to delineate the ethological peculiarities of the same and other species, trying to fix some general characteristics of the genera and of the subfamily.

3.2 Ethological Characters of Hover Wasps

In Chap. 5 of their book on "Bionomics of social wasps and bees in equatorial Sumatra" Soichi F Sakagami and Soichi Yamane (1990) deal with the behaviour inventory of two species of hover wasps studied on the Indonesian island. The authors begin by stressing the fact that previous students of these wasps paid particular attention only to social behaviour or to nest architecture, as the result of complex behavioural sequences, while most "basic" patterns (with the exception of some noteworthy behaviours distinguishing these wasps from other social companions, such as the collection of abdominal substance or egg deposition) were practically ignored. The Japanese entomologists, in my opinion, rightly observed that useful behavioural characters for establishing the phylogenetic relationships with other social wasps can only be those that are shared by all groups. They also stressed the fact that ethological characters are "inevitably less precise" (than morphological and biomolecular ones) and "only available for limited taxa" to be considered in the reconstruction of evolutionary history. I can object, in any case, that good descriptions of basic behavioural patterns are really lacking for other social wasps with the exception of some species. I can give the example of the genus *Polistes*, which is certainly the most studied of all the Polistinae, where we know the behavioural inventory of several species from Eurasia and North and South America. To the best of my knowledge, comparisons between different species of the same or different subgenera are quite rare, even if some behavioural patterns seem to vary consistently. One of the reasons is that, unlike nest architecture (that resumes a set of coordinated behaviours in fixed and measurable objects), descriptions of most behavioural patterns are so difficult and incomplete that even if accompanied by drawings or pictures, they are totally useless for making comparisons not only between different taxa but at the intra-specific level as well. One partial solution to this problem could be to collect video-sequences on at least the principal behaviours of a species and begin to organise video data banks for different taxa using the facilities offered by the web and social networks. To my knowledge, so far this is only available for some *Polistes* wasps on the Joan Strassmann and Dave Queller research group web site: http://www.joanstrassmann.org/Joan_Strassmann/Polistes.html).

This chapter intends to partly fill this gap for Hover wasps by commencing to list and describe the principal behaviours of some species with the visual aid of videoclips published online.

Table 3.1 Categories of main behavioural patterns in females and males on and off the nests

	Elementary behaviours	Colony maintenance behaviours	Social behaviours
On the nest	Resting	Defence from predators and parasites	Defence against conspecifics
	Self-grooming	oviposition	Adult–adult trophallaxis
	Self-feeding	Larval care	Dominance-subordinance
		Nest construction	
Outside the nest	Resting	Hunting	Male clustering
	Self-grooming	Nest material collection	Mating and related behaviours
	Self-feeding	Sugar and water collection	
	Flight		
	Defecation		

Sakagami and Yamane divide the behaviours of *P. mellyi* into three functional categories (1) Self-maintenance behaviours; (2) Colony maintenance behaviours and (3) inter-adult (social) behaviours. I agree with the general organisation of the matter but disagree with the treatment of some behavioural patterns. I believe we can simplify the overall description of the behaviours and propose the scheme in Table 3.1, which can be used both for males and females, distinguishing activities performed on and off the nest.

3.3 Presence On and Absence Off the Nest

In several species, maximum presence of marked individuals on the nest in a given colony (individuals accepted with no signs of hostility) usually occurs just after sunset.

In *L. vechti*, the number of wasps on the nests (recorded for 28 colonies) after sunrise falls to 60 % of the population recorded and remains almost constant all day long. The percentage rises gradually at dusk but never reaches 100 % because some individuals do not come back at night (Turillazzi 1990a). Presence on the nest in *P. jacobsoni* is highest at dawn and dusk but also shows a lower peak at midday (Turillazzi 1988). For *E. calyptodoma*, Hansell (1987a) gives an in/out traffic of females to and from the nest that is greater during the mid part of the day. In *L. topographica* Baracchi et al. (2009) found that the two peaks in off-nest activity of colony members at late morning and late afternoon, observed over an entire day, were consistent with reports for other hover wasps (Fig. 3.1).

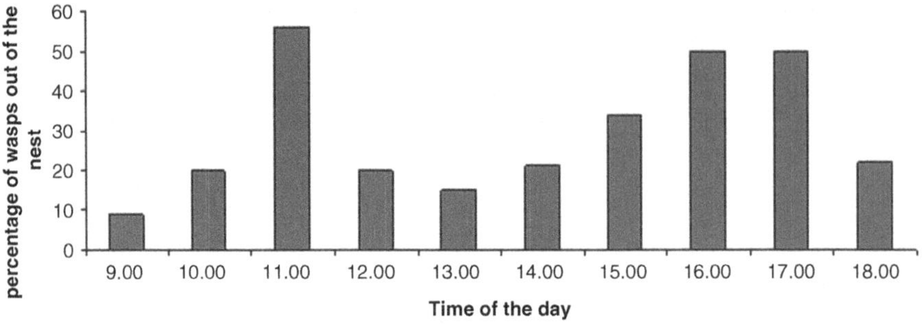

Fig. 3.1 Percentage of adult wasps away from a nest of *Liostenogaster topographica* (based on maximum number of marked wasps in the colony) during one day observation (from Baracchi et al. 2009, redrawn)

3.4 Elementary Behaviours

I include here some of the behaviours proposed by Sakagami and Yamane (1990) and in particular: resting, self-grooming, self-feeding, flight, defecation and mating, which are common to females and males and constitute the principal "activities" of the adult inhabitants in a colony. The first three can be observed both on and off the nest, while the second three are performed only away from the nest. Only slight differences in the various activities are evident between sexes, species and genera and some (such as defecation and copulation) have been observed only in certain species.

3.4.1 Resting

It seems that the principal activity of all hover wasps is complete inactivity. Actually these insects are quite boring to observe compared with other social wasps, which does not mean that their behaviour is simpler to describe. This is due to the fact that long periods of resting are followed by short and often unpredictable intense bursts of frenetic activity, when all the individuals of a colony begin "to do something", that stops just as suddenly as it began.

Resting positions of males and females in all the species are fundamentally the same: in *Parischnogaster* there is the tendency for the individuals to keep their body completely distended. Sakagami and Yamane (1990) distinguish various positions of resting in *P. mellyi* and confirm the observation of Hansell (1983) that the upper resting position on the nest is occupied more frequently by the dominant female. Individuals of the genus *Liostenogaster* usually keep their gastra slightly bent ventrally or laterally (Samuel 1987; Turillazzi 1990a); in *Eustenogaster* females keep their gastra distended when on guard at the nest entrance (ready to produce the beating sound—see Fig. 5.4) and more often

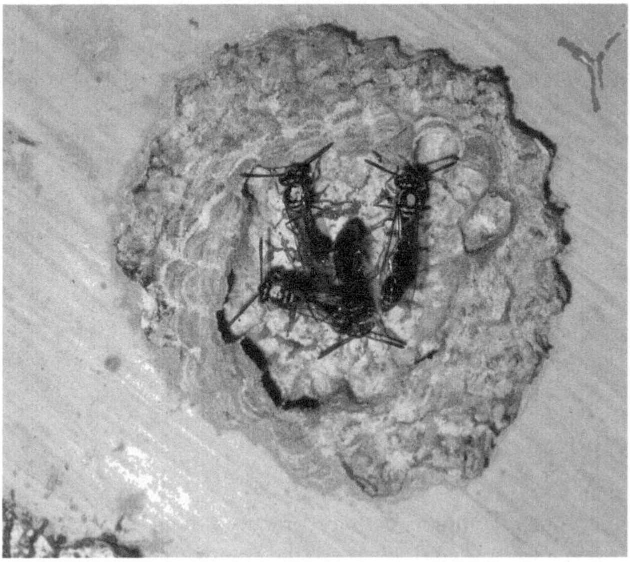

Fig. 3.2 Resting positions of *Liostenogaster vechti*

(but see the straight positions on the combs of *E. calyptodoma* reported by Hansell 1987) bent when on the comb.

In the round nests of *L. vechti*, females and males can arrange themselves in the central area of the nest, with their heads facing the cell openings and external environment; their position is probably related to active defence against possible predators, although it does not appear that particular individuals maintain special positions (Fig. 3.2).

In *L. flavolineata* Samuel (1987) observes that the wasps (more commonly the males) sometimes rest with their gasters or heads in the cells. In this species, and also in *Parischnogaster*, resting is occasionally accompanied by raising and lowering the wings and gentle slaps on the abdomen (which could be interpreted as anti-parasitoid behaviours).

Female and male *Metischnogaster* usually rest deep in the cells; this position is possible by the rather long cells which are peculiar to this genus (Turillazzi 1990b). According to Spradbery (1975), in *Stenogaster concinna*, the females also spend most of their time inside the long cells of their nests.

In *P. nigricans serrei* both females and males spend over 50 % of their time resting with their bodies slightly raised or flattened against the nest, their metasoma extended or slightly bent, wings lowered and antennae stretched apart (Turillazzi and Pardi 1982). In this species individuals never rest inside the cells but can often take a break between the cells and ant guard.

In *P. jacobsoni* I found that adults tend to rest on certain sections of the nests, especially in relation to cells holding particular stages of the immature brood. In one colony observed for over 5 h, four females were present on certain defined segments of the nest whilst the dominant female usually rested on two operculated cells in a section that the other females hardly frequented at all (Turillazzi 1988). In *P. nigricans serrei,* on the other hand, a dominant female was observed to rest mainly on the central cells of the nest and on the substrate near the ant guard.

In various species, females may also rest in the vicinity of the nest: in nest clustering species *Liostenogaster flavolineata* and *L. vechti* various floaters (or vagrant) females can be found during the day, but especially at night, resting in the spaces between nests (Samuel 1987; Turillazzi 1990a; Field et al. 1998; Coster-Longman 1998).

In *P. jacobsoni*, individuals rest during the day especially at dawn and dusk but also in the middle of the day (Turillazzi 1988).

3.4.2 Self-grooming

This is another quite common activity practised both by males and females. In *Parischnogaster mellyi* and *P. alternata* the complete sequence of self-grooming is too complex and elaborated to be conveniently described and I suggest giving a look at the Videoclip 3.2 published online. Grooming usually begins with movements of the hind legs which rub together followed by the abdomen and the wings. The wings are rubbed against each other and against the dorsal and ventral part of the gaster, then the hind legs are rubbed against one of the middle legs which then continue to rub against the fore legs and mouth parts, while the wasp stays on the other three legs. The fore legs then begin to rub the head, mouth parts and antennae. The sequence is almost the same for males and females and basically similar in other genera (see Sakagami and Yamane 1990 for *L. vechti*). If we look at the self-grooming sequence and the behaviour with which the females of some species of *Parischnogaster* (*P. nigricans serrei* and *P. jacobsoni*) apply the Dufour's gland secretion to the nest ant guard, we realise that they are fundamentally similar and that the movements of the legs are the same. In ant guard construction, a small ball of secretion appears at the tip of the gaster which is then collected by the movements of the hind legs and transported via one of the middle legs to the mouth parts to be applied on the nest substratum (see Fig. 6.64) The only difference with self-grooming is that there are no rubbing movements over the entire body in the ant guard application procedure. This made us think that during the self-grooming manoeuvre, substances coming from the gaster can be applied all over the body (Turillazzi and Pardi 1981) and this is also the impression we had looking at slow motion video recordings. I have to say that cuticular hydrocarbons do not differ much in composition from those found in the Dufour's gland (see Fig. 6.66); moreover recent analyses found that venom peptides also occur on the female cuticle. This is evidence that these secretions can be smeared over the body with grooming movements. The slight antimicrobial activity of venom peptides reported in various species of hover wasps induced Baracchi et al. (2011) to suggest that this habit for the females represents a sort of additional protection against possible pathogens besides a simple mechanical action for eliminating dust and particles from the body.

3.4.3 *Self-feeding*

Adult females and males feed especially on sugar sources which can be various (in captivity they promptly accept any kind of sugary solutions, including water with sugar or honey). Prey collected for the larvae are masticated for a long time before being supplied to the brood. Both males and females also feed on the food already given to the larvae and can even chew the abdominal secretion collected on eggs and larvae for a while before discarding it (this can easily be verified by placing a black plastic sheet under an active colony: after some days this will be covered with tiny bits of abdominal secretion). Adults can also feed on eggs or small larvae captured on visited nests of their own or other species (see Turillazzi et al. 1997).

3.4.4 *Flight*

Special flight characteristics confer to these wasps the name of hover wasps. I remember that in West Java village people call these insects by the dialect name of "papantin", which clearly derives from "papaton", used for indicating dragonflies. Especially in *Parischnogaster*, *Metischnogaster* and *Eustenogaster* flight is characterised by complex hovering and deft and rapid changes in flight direction. This high manoeuvrability allows females to collect small prey from spider webs and males to perform their aerial displays (see later). Hovering species keep their legs perfectly close to their body and their abdomen fully distended (even if they can still bend it when necessary). On the contrary, species of *Liostenogaster* have a less manoeuvred flight and usually keep their legs hanging down. The charming flight of species of *Parischnogaster* can be better appreciated in the Videoclip 3.1.

3.4.5 *Defecation*

In *Polistes* wasps, individuals can often be seen defecating from the nest, after extending their abdomen backward and emitting the faeces which fall down without touching the comb. A similar behaviour has never been observed in any species of hover wasp and I was only able to see how these insects defecate by following captive individuals. In fact hover wasps defecate away from the nest and in *P. mellyi* both females and males stop on the edge of a leaf, or other substrata, and drag the tip of their bent abdomen over it, leaving behind hindgut content on the spot. Beani et al. (2002), observing a captive population, discovered that both males and females defecate at specific places and that this activity has an important part in marking specific "hotspots" in the jungle where mating takes place (see following paragraphs and Sect. 5.3.2.3).

3.4.6 *Mating and Related Behaviours*

Mating systems in social wasps have not been extensively studied in depth, with the exception of some species of *Polistes* wasps: our group performed various studies on the matter with contributions from various researchers (for a review see Beani 1996). Hover wasps have also been the object of intense studies by our group, which examined both the morpho-functional apparatuses of the males and their sexual behaviours.

3.4.6.1 Male Aerial Patrolling

Observations on the behaviour of Stenogastrinae males began with H.T. Pagden who was the first to describe the particular patrolling behaviour of males of *Metischnogaster cilipennis* and *M. drewseni* (Pagden 1962). These hover like dragonflies in shady wet places of the jungle during certain hours of the day, contracting and extending their abdomen and revealing large white stripes on the anterior part of their gastral tergites, which are almost invisible in the resting position. These insects are truly astonishing because their aerial display is so evident against the dark scenario of the jungle. They perform in forest clearings or over swiftly running streams; in fact, when they extend their abdomens they look as if they are switching on little lamps. Pagden noted that the display became particularly evident when other males visited the patrolling spots and hovered back and forth behind the patrollers. In the same paper Pagden briefly reported other interesting observations and also suggested that in the males of the two species of *Metischnogaster* extending the gaster could trigger pheromone release. I had the opportunity of observing patrolling flights of male *M. drewseni* on several occasions and was enchanted to watch these little insects dancing above the flowing stream in the late afternoon (Fig. 3.3). Angelo Fortunato and Christina Coster-Longman noted that patrolling males change flight direction so they are always facing the wind, even if slightly, and confirmed this tendency using an electric fan to influence the wasp flight (Fortunato and Coster-Longman 2000; Fortunato unpublished observations).

Pagden observed that males of species of *Parischnogaster* also perform hovering abdominal white stripe displays and reported a mating that occurred on a leaf between a female and a patrolling *P. striatula* male (Pagden 1962). During the second mission to Indonesia I had the opportunity to observe male patrolling in two species of *Parischnogaster* (*P. mellyi* and *P. nigricans serrei*) which in some cases shared the same sites to perform their flights (even if the second preferred wider, better lit areas). It soon became evident that the males of the two species showed certain small differences in their behaviour and had different patrolling times. When a male *P. mellyi* hovers, it faces a landmark, usually a leaf or a twig, for a number of seconds, afterwards it makes a long detour, usually horizontal, to the right or left and then returns to the landmark. When hovering at the landmark, the wasps can extend

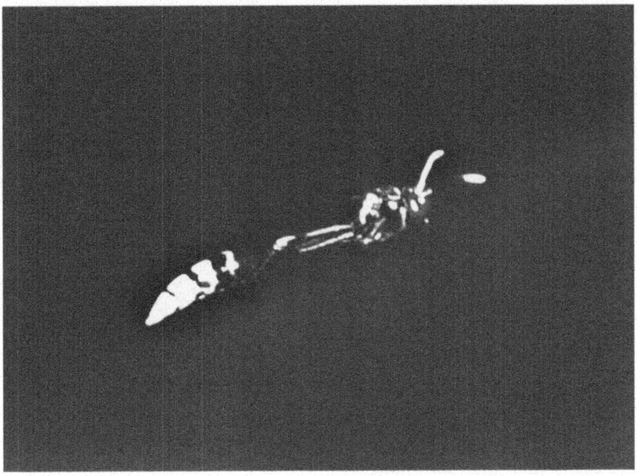

Fig. 3.3 A male *Metischnogaster* during aerial patrolling and gaster display (Photo by I. Petrocelli)

and contract their gasters to reveal the white stripes on their metasomal tergites. Beani and Turillazzi (1999) recognise at least seven behavioural categories of male interactions during patrol flights of *P. mellyi*, going from solitary displays with the white dorsal stripes manifested for 5–40 s, to actual demonstrations of strength consisting in aerial skirmishes between two rivals accompanied by extreme white stripe displays. Fights can degenerate into true attacks performed by the two contenders, hitting each other with the legs or abdomen.

The behaviour of *P. nigricans serrei* males differs slightly because their hovering flights include some vertical shifts, fewer horizontal flights and fewer abdominal stripe displays (Turillazzi 1983a). A single male can switch between different landmarks which it contends with other patrollers. If a landmark is already occupied by another male, the newcomer begins to hover a few centimetres behind and the owner usually reacts swiftly by extending its gaster. Following this display the newcomer (and sometimes the owner) leaves the landmark unless the encounter breaks into a contest. This consists in a tandem flight of the two males which begin to ascend, suddenly changing their positions (the male in front goes behind and vice-versa), mutual stripe-displays and eventual flying clashes, until one of the two leaves and the other returns to hover at the landmark. At the end of the daily period of patrolling, which can last for over an hour, fewer and fewer males can be observed at the landmarks and, usually, the last males to remain are those which have won the most contests (Turillazzi 1983a) (this was confirmed by observations in a flying room, Beani and Turillazzi 1999).

Patrolling flights take place at well-defined hours of the day. In West Java *P. mellyi* leave their nests or clusters "en masse" around 11.00 AM. and come back around 12.45, while massive departure from the nests of male *P. nigricans serrei* occurs around 1.00 PM and the patrols end 2 h later.

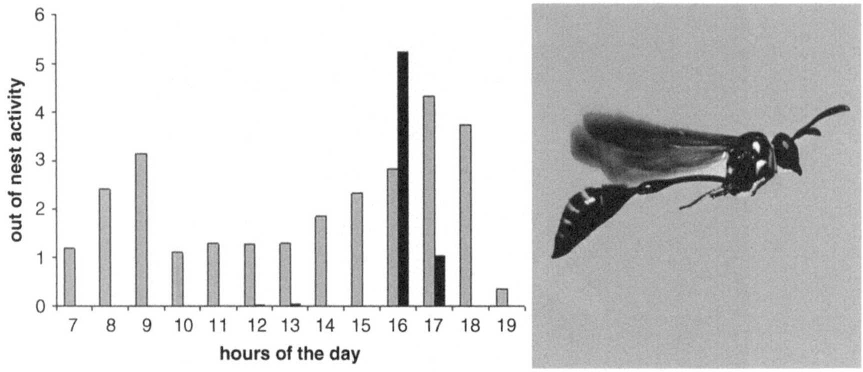

Fig. 3.4 *Left*: daily "outs" of males (black columns) and females (grey columns) of *Parischnogaster alternata* from a cluster of nests placed in a hut at Genting Tea Estate (Pahang Malaysia). *Right*: stripe display of a male *Parischnogaster alternata*

In Peninsular Malaysia the patrolling period of *P. mellyi* occurs roughly at the same time as in Java while that of males of *P. alternata* begins around 3.30 PM (Turillazzi and Francescato 1989; L. Pizzocaro and G. Cusseau, unpublished) (Fig. 3.4). The timing of patrolling behaviour is probably determined by an internal clock adjusted on the photoperiod as individuals reared in captivity conform their patrolling activity to the L–D periods imposed in the cages.

Elisabetta Francescato and I had the opportunity to observe and describe the patrolling behaviour of males of *Liostenogaster flavolineata* and of *Eustenogaster* sp.; the first were observed in a captive population reared in a flying room, the second were observed in the field (Turillazzi and Francescato 1990). Males of *L. flavolineata* performed rapid and almost linear flights between various perching points where they rested for some minutes. Perching points were usually visited following a similar sequence and were practised by different individuals. Generally after arriving on their perches the males bent their gastra ventrally and performed linear walks of some centimetres dragging their gastral tergites over the substratum (Fig. 3.5).

If two males visited the same perch at the same time, the one already on the perch would dart forwards towards the other while the latter was still hovering before landing. The patrolling activity of males recorded in the laboratory occurred between 4.00–5.00 PM and 5.00–6.00 PM. and roughly coincided with the period of absence from the nest noted in the field for males of the same species Samuel (1987). The dragging activity of the males is evidently related to the release of secretion from clusters of unicellular glands associated with the intersegmental membranes between the 3rd, 4th, 5th and 6th gastral tergites and with conspicuous transverse bulges (which we called scrapers) on the 3rd, 4th, 5th and 6th gastral tergites (see Fig. 2.21).

The patrolling behaviour of *Eustenogaster* males had first been noticed by Williams (1919) who mentioned the aerial exhibitions of a male hovering repeatedly over the same leaf. We observed a similar behaviour in a male of

Fig. 3.5 A male *Liostenogaster flavolineata* dragging its abdomen over a perching leaf (Turillazzi and Francescato 1990)

Eustenogaster (prob. *micans*) which landed several times on two nearby leaves, passing from one to the other with short flights. When on the leaf, the male remained motionless with its abdomen raised and its wings and antennae held raised and slightly apart. When another male arrived on the spot, an aerial duel began and lasted until one left. Patrolling took place at the same site starting at about 10.30 AM and finishing about 1 h later. At other sites I observed males, probably of another species, performing similar flights but around 1.00 PM. Up to six males were engaged at the same time on a perching leaf, but without any apparent display. Males of *E. fraterna* possess a band of unicellular glands on the anterior end of the 3rd gastral tergite, the outlets of which occur in a tegumental area that seems to be exposed when the wasps are in the perching position (Turillazzi and Francescato 1990) (Fig. 2.21).

3.4.6.2 Mating

Two papers by Beani and Turillazzi (1999) and by Beani et al. (2002) again examined patrolling behaviour of *P. mellyi* in captivity and clarified that it is especially related to the mating system of the species (see also Sect. 5.3.2.3). During various months of observation in a flying room it was ascertained that (1) males are usually active for 3 weeks or more, up to a maximum of 32 days, (2) males seldom land during hovering, and they do this more frequently towards the end of patrolling, walking along the

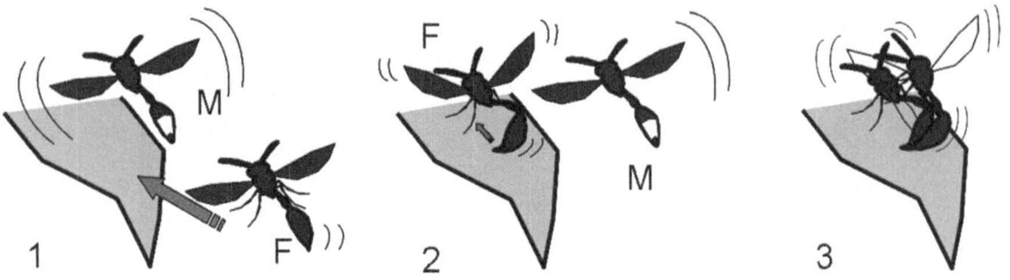

Fig. 3.6 A sketch sequence of mating in *Parischnogaster mellyi* (M:male; F:female) (see text)

edge of leaves or paper strips, dotting the substrate with the tip of their slightly bent abdomens, (3) at the end of activity males feed themselves at sugar sources and then return to nests or to communal resting points. On the other hand it was seen that some females usually appear at the patrolling sites especially during the second hour of male activity, having sexual interactions.

In a typical interaction the female approaches a displaying male from behind, extending her legs towards the male gaster. The male prolongs the aerial display often performing zigzag movements (Fig. 3.6-1). At this point the female lands on the hover site and repeatedly touches the substratum with the tip of her gaster (Fig. 3.6-2). The male rapidly touches the female's back with his legs, grasps her thorax for a few seconds and performs genital linkage with the linked abdomens fully extended for 2–25 s (Fig. 3.6-3). Afterwards, the female will sometimes clean her genital area with her legs (Beani and Turillazzi 1999).

Some years before I had already observed males of *P. mellyi* in captivity attempting homosexual mating; they were on a leaf used as a patrolling landmark, in a particular position that I reported in a drawing (Turillazzi 1983b).

Observations and experiments in a captive population led us (Beani and Turillazzi 1999) to conclude that the patrolling and stripe displays of male of *P. mellyi* (and other species of *Parischnogaster*) represent an example of an aerial lek system and a status badge display that is comparable to those of some vertebrates (Brown and Macdonald 1985). Sakagami and Yamane (1990) report about another possible mating of *P. mellyi* observed in the field. This also occurred during the period of male patrolling, precisely in a patrolling area along a pillar of the Minankabau Royal Palace in Sumatra.

3.5 Colony Maintenance Behaviours

3.5.1 Defence from Predators and Parasites

Active defence of the colony is supplemental to passive defence, more practised in hover wasps that relies on nest camouflage, small colony size and repellent barriers. Stenogastrinae do not have the weapons to successfully counteract large predators (see Chap. 2) and are defenceless against the attacks of hornets that are their most dreadful enemies. Escape is the main form of self-defence that the adults have against possible large predators; but this can indirectly protect the nest and immature brood as well, as it creates a confusion effect for the predator. In *Parischnogaster*, individuals fall from their nest when under attack to resume flying only near the ground: in this way they make it more difficult for a possible predator to discover their nest (Landi et al. 1998). *Liostenogaster* usually escape directly flying in all directions from their nest or, in the case of cave nesting species, towards the light, sometimes using their stings to hit a large intruder. *L. abstrusa* on the contrary, which nest inside hollow dead branches, hardly abandon their home at all, preferring to remain hidden and protected: if they are chased from their nest they suddenly come back and hide again while other species can stay several minutes away from their nests (personal observation). Other species can perform active defence even against large predators: I'm speaking about some species of the genus *Eustenogaster* which are the largest of all the hover wasps. Chris Starr furnished personal impressions about the sting of female and male *Eustenogaster* in his commentaries on wasp collections in the Philippines (Starr 1984) but it is not clear if he also witnessed species of this genus defending their nests. Starr observed that the peculiar efficiency with which male *Eustenogaster* react, stinging any possible predator with their pointed parameres, would favour both females and males of the same species. For my own part, I can confirm that I was once the object of repeated attacks by females from a large, and full of immature brood, colony of *E. hauxwellii* which stung my camera while I was trying to take some pictures of their colony. I tried, in vain, to obtain the same reaction from colonies of the same species in other circumstances; evidently the status of the colony itself is important for triggering the aggressive defence of the inhabitants.

Hover wasps are able to defend their colonies from small-sized predators using particular behaviours. Ants are the principal enemies of wasps especially in tropical climates (Jeanne 1979). Hover wasps build their nests in places where the presence of ants is not so high but, in the case of scout ants arriving near their colonies, they can promptly react by assuming a defence position which consists in presenting the tip of their laterally bent abdomen to the intruder (gaster bending). The manoeuvre is often accompanied by wing buzzing and darting with open mandibles; small ants are bitten and thrown from the nest while larger ones are slapped with the back of the gaster. The wasps can then grab them and then drop from the nest taking the ants with them. Other small intruders on the nest are dealt with in a similar way.

Flying "things" approaching the nest with uncertain flight are followed with interest by resident wasps as soon as they are detected. If the object is smaller than a certain size, resident females can begin to buzz their wings and then start to hit it mid-air. This kind of defence is particularly used by females of *P. mellyi*, *P. nigricans serrei*, *P. jacobsoni* and *L. vechti*, against intruder conspecifics. Episodes like these can be extremely frequent and I recorded dozens of them in a short time in a cluster of *L. vechti* colonies on the return to the nests of a large number of individuals due to sudden heavy rain. It is not just the females which present this reaction as I observed it in two males of the same species.

Experiments performed in the laboratory with the presentation of various types of cues to colonies of *P. mellyi* showed that intensity of response is correlated to the conditions of the immature brood and to the number of adults present in a colony (Landi et al. 1998)

In *L. flavolineata* Samuel (1987) reports about 134 instances of females repelling foreign conspecifics in a nest observed for 102 h. Resident females rush to the edge of the nest with their antennae raised and pointing forward against any intruder approaching the nest with an uncertain, hovering flight. They can begin to flap their wings, buzzing and flexing their gasters. Even a male, in the absence of the only resident female, was observed in one case defending its nest with its gaster bent against a foreign female conspecific.

In *Anischnogaster* we observed unusually long sequences of nest patrolling (a behaviour that in other genera commonly follows the discovery of intruders in the nest) performed by females of *Anischnogaster* sp. A and sp. B. These can be interpreted as careful and prolonged checks of Tachinid parasitoids that are one of the main problems for the immature brood in species of this genus (Turillazzi and Hansell 1991). Again, the reduced colony size found in the species of this genus has been interpreted as a response to the high pressure exerted by these parasitoids (Turillazzi and Hansell 1991).

3.5.2 Collection of Abdominal Secretion and Oviposition

In Chap. 2 I had already mentioned the importance of the secretion of the Dufour's gland in stenogastrine biology; here I refer to the collection of the secretion during larval care and the oviposition process observed in particular in species of *Parischnogaster* and *Liostenogaster*.

When collecting the secretion, the wasp stays with her hind and middle legs on the nest and the gaster bent ventrally towards the mouth. The anterior legs are kept slightly bent and gently touch and roll the secretion into a bolus that appears at the tip of the gaster, with the distal part of the tibiae and proximal part of basitarsi. The gaster performs back and forth movements towards the mouth while the tip seems to trace small circles on the sagittal plane. The head and front legs follow the abdominal movements. The movements are more frequent at the beginning of collection and slow down towards the end as the bolus increases in size (Fig. 3.7). Duration and average pumping rates per second of substance collection can vary in different species. As soon as the collection of substance is finished the bolus is placed in the cell (Turillazzi and Hansell 1991) (see Videoclips 3.3 and 3.4).

Fig. 3.7 A female *Parischnogaster striatula* collecting abdominal substance (photo by D. Baracchi)

Oviposition has been directly observed in various species of four genera: *Parischnogaster* (*P. mellyi*; Hansell 1982; Sakagami and Yamane 1983; Turillazzi 1985a; *P. nigricans serrei*. Turillazzi and Pardi 1982; *P. striatula, P. alternata, P. jacobsoni*, Turillazzi 1985b), *Liostenogaster* (*L. flavolineata,* Samuel 1987; personal observation; *L. vechti*, Turillazzi 1990a), *Eustenogaster* (*E. calyptodoma*, Hansell 1987; *E. fraterna*, Francescato et al. 2002) and *Anischnogaster* (*A.* sp. Turillazzi and Hansell 1991) and is fundamentally similar in all the hover wasps. The characteristics of laid eggs reported in species of another two genera (*Metischnogaster,* Turillazzi 1990b and *Stenogaster*, Spradbery 1975) suggest that the basic behavioural sequence is also valid for these groups.

Egg laying in hover wasps is peculiar (Fig. 3.8). In all the observed species of three genera (*Parischnogaster*, *Liostenogaster*, *Eustenogaster*) (Hansell 1982, 1987; Turillazzi and Pardi 1982; Turillazzi 1985a; Samuel 1987) it consists of three phases: after initial inspection of a cell, the female bends her gaster ventrally towards her mouth parts and collects a patch of the abdominal substance produced in the Dufour's gland. Then the wasps may inspect the cell again, retaining all the substance in the mouthparts. After stretching the gaster the wasp bends it again and opens it at the tip with the sting extruded dorsally, towards her mouth. After some time the wasp collects the eggs as it emerges allowing its concave surface to adhere to the patch of abdominal secretion (see Videoclip 3.5). The egg is then placed in the cell and stuck to the bottom by means of another small patch of secretion (once thought to be

Fig. 3.8 A female *Parischnogaster striatula* during phases of egg deposition. (**a**, **b**, **c**) The female is attaching the egg to the pad of Dufour's gland secretion it holds in her mouthparts. (**d**) The female is going to place the egg inside the cell (photo by D. Baracchi)

different from the Dufour's gland one, given their different density and colour some hours after deposition—Turillazzi 1985b) present on its convex surface.[1]

[1] In my description of this behaviour in five species of *Parischnogaster* (Turillazzi 1985b), I stressed the fact that, probably, the first bolus of abdominal secretion did not serve to attach the egg to the bottom of the cell. The main reason for this is that the egg is already furnished with a drop of secretion when it emerges from the tip of the abdomen. If we check laid eggs it is clear that the consistency and colour of the secretion which glue them to the bottom of the cells are somewhat different from those of the Dufour's gland secretion. However Sakagami and Yamane (1990) contested this and proposed that in any case the drop visible on the convex part of the egg is in fact the same of that collected in the mouth by the wasp. I thought that the only way to clear this was to perform chemical analyses of the two substances and recently, helped by Luca Calamai of the Centre of Mass Spectrometry of University of Florence (C.I.S.M.), I performed Gas Chromatographic analysis obtaining almost identical spectra for the secretion placed on a recently laid egg of *P. mellyi* and for the small amount of secretion on which the egg was glued to the bottom of

The first patch of abdominal substance is left on the concave surface of the egg. After placing the egg, the wasp bends her gaster again and collects a larger patch of abdominal secretion that is added to the previous patch on the egg. In *P. striatula* on two occasions I observed wasps, disturbed by my movements during the deposition process, to abandon the nest with the eggs in their mouthparts and return after a while to place the egg in the cell and then collect the second patch of abdominal substance.

In *Eustenogaster* oviposition has been reported by Hansell (1987a) in *E. calyptodoma* and by Francescato et al. (2002) in *E. fraterna*. In the second species we found that in the three observed depositions the second collection of abdominal secretion (phase three) was missing. This also explains why some eggs have very little secretion or lack it all together (Francescato et al. 2002).

As regards *Stenogaster*, egg laying has never been directly observed in the only species studied so far, *S. concinna*, but Spradbery (1975) reports to have found secretion only on some of the eggs he observed in various nests. This could indicate that the second addition of secretion is possibly delayed and not performed in connection with egg deposition.

Also *Anischnogaster* presents interesting differences with the other genera. In *Anischnogaster* sp. A, Mike Hansell and I observed oviposition only on two occasions, but in one case there was no collection of the second bolus of abdominal secretion, while in the other egg deposition was neither preceded nor followed by collection of secretion. In this species we observed secretion on five of the seven eggs found in three nests while in another undetermined species (*Anischnogaster* sp. B) secretion was present on all the five eggs we found (Turillazzi and Hansell 1991). J.P. Spradbery (1989) reports that in *A. iridipennis* he found secretion only on some of the eggs he examined. In *A. laticeps*, however, we did not find any trace of secretion on any of the 26 eggs nor on the small larvae we found in 25 nests. This raises the question of the actual function of this substance in this genus, as the size of Dufour's gland in *A. laticeps* does not differ much from that of the other genera, even though the ratio between average gland length and head width is the lowest (1.58) with respect to those of species of other genera (*L. vechti*, 1.67; *Eustenogaster sp.*, 1.71; *P. striatula*, 1.75; *L. flavolineata* 1.81) (Turillazzi and Hansell 1991). We only found a positive correlation between the size of the Dufour's gland and ovarian development; this could suggest that the secretion, in this species, serves solely for gluing the eggs to the bottom of the cell and that the larva remains in situ by in some way clinging to the egg shell, as occurs in other social wasps.

Indirect deposition of the egg is absent in other social Hymenoptera, with the exception of some ants, and there is still a mystery about the possible steps that led to the evolution of this behavioural pattern. One possible hypothesis is the following (1) Primitive direct oviposition (as occurs in other social wasps) with the eggs

---

the cell: this undoubtedly confirms the hypothesis of Sakagami and Yamane that the substance that is present on the convex part of the egg just when it emerges from the gaster is also completely, or for the most part, produced by the Dufour's gland (see also Fortunato and Turillazzi 2012).

placed directly on the bottom of the cells attached with a drop of Dufour's gland secretion already present on the convex walls. (2) Addition of a further patch of Dufour's gland secretion (this time collected by the female in her mouth to be more precise in applying it on the already laid egg) would protect the eggs from predators such as ants. (3) The first and second phase could have blended so that the patch of secretion also became a device for holding the eggs without the risk of breaking them. (4) A substantial amount of abdominal secretion placed on the eggs could have become a suitable substrate for rearing the small larvae, furnishing protection and a place where to store small amounts of food (see also Chap. 2 and Fortunato and Turillazzi 2012, for morpho-functional adaptations).

3.5.3 Larval Care

Cell inspection is one of the most common behaviours of females and males. Inspection can last just a few seconds when the brood are simply checked, but it is more thorough when the females deposit food on the secretion (in the case of small larvae) or inside the coils of more mature larvae. Adults will often retrieve boluses of food or fluids from the larvae in order to nourish themselves or to transfer them to other brood (Turillazzi 1985c). A set of interactions between adult females and larvae are presented in the Videoclip 3.6. In the clip, it is evident how the coiled larvae react to the stimulation of the mandibles of the adults, as they open up like sphincters: the same reaction can be obtained experimentally by stimulating the pleural lobes of the coiled larvae with a slender stick or a glass capillary tube (Fig. 3.9).

Larval care also includes the elimination of larval faeces collected in the peritrophic sacs of the pupating larvae. In species which do not close the cells of the pupating larvae the peritrophic sac, resembling rosary beads (or a single sausage-like sac as in *Metischnogaster*), is simply collected by the adults as it is emitted by the larvae and then discharged by dropping it from the nest. In species that close the cells, (such as *Parischnogaster mellyi* or *P. nigricans serrei*) the females open them after a few days to eliminate the peritrophic sac and then seal the cells again with material generally taken from the nest itself. This can be shown in a practical way by colouring the opercula of newly sealed cells which will be found with a different colour patch after some days (the same can be observed in the nest of *Liostenogaster filicis* of Fig. 6.31) (Turillazzi and Pardi 1982).

In *Liostenogaster flavolineata* it seems that the adults do not eliminate the larval meconia straight after pupation but only after the emergence of the brood (Samuel 1987).

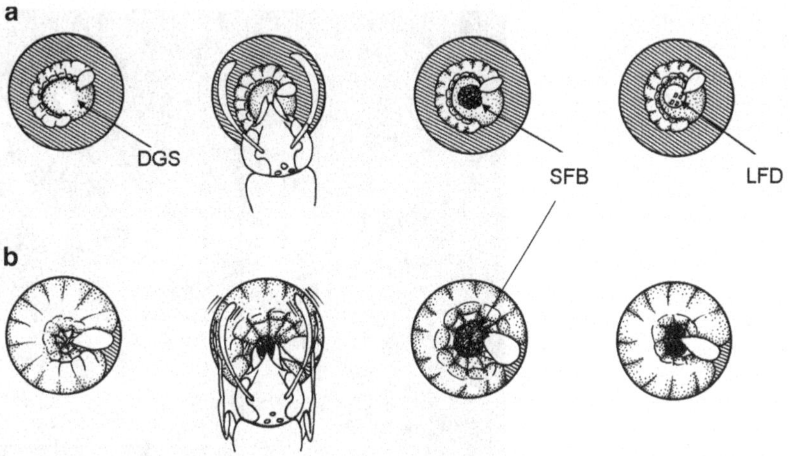

Fig. 3.9 Schematic representation of feeding the larvae. (**a**) Small larva, (**b**) more developed larva (from Turillazzi 1990c, re-drawn). *DGS* Dufour's gland secretion; *SFB* solid food bolus; *LFD* liquid food drop

3.5.4 Foraging

The peculiar behaviour of some hover wasps plucking small prey in flight from spider webs was reported by the very first authors (Williams 1919; Pagden 1958; Iwata 1967). I have personally observed species of the genus *Parischnogaster*, *Eustenogaster* and *Metischnogaster* performing this kind of foraging, which is obviously possible thanks to the manoeuvred hovering flight of these species. In *P. nigricans serrei* the females easily hover close to the webs without touching the threads and pick off the prey with their legs and mandibles (Fig. 3.10, see also Videoclip 3.1) (this particular ability can be exploited when rearing these wasps in captivity, as they can easily take prey if it is attached on vertical plates placed in the cages). Females may even attempt more than one grab in the same foraging flight and try to detach small prey from the spider web using all their legs; if the prey is too big they cut it with their mandibles. Treatment of the prey begins in flight, just after its capture and the foragers can spend some time chewing the bolus before returning to the nest. In nature a single wasp can return with food three or four times in a few minutes.

This kind of foraging is certainly practised by those species capable of the most efficient hovering flight but it has never been observed in species of *Liostenogaster* which, on the contrary, have a less manoeuvred flight. However I have often observed *P. mellyi* and *P. nigricans serrei* hovering in the midst of swarms of tiny midges in sunny patches in the forest, extending their legs and trying to catch the small insects. According to a report by Samuel (1987) *L. flavolineata* can perform this kind of hunting as foragers of this species return to the nest with winged termites, probably caught during swarming. The use of the sting has never been observed during prey capture.

Fig. 3.10 A female of *Parischnogaster alternata* plucking a small prey from a spider web (photo by D. Baracchi)

Sugar sources can vary and I have often seen females and males of *Parischnogaster* feeding on plant sap (gen. *Manyot*) or licking the surface of leaves spotted with honeydew. Hover wasps can also visit flowers for nectar. Another possible source of solid and liquid food can be the nests of conspecifics and congenerics where females can prey on eggs and small larvae or steal abdominal secretion and solid and liquid food from the larger ones. In a mixed cluster of colonies of *P. alternata* and *L. flavolineata* we often observed females of the latter species arriving on the unguarded nests of the former to perform stealing activities (Turillazzi et al. 1997).

In *P. nigricans serrei* in Java foraging activity reaches a peak in midmorning, decreases in the afternoon and then rises again before the end of off-the-nest activities (Turillazzi 1983b), while Pagden reports a similar daily activity for *Metischnogaster cilipennis* in Malaysia (Pagden 1958).

The extent of home range for these wasps has not yet been clearly established for any species. Capture–release experiments were performed by myself (Turillazzi 1983b) on *Parischnogaster nigricans serrei* and, recently, by some students of mine (Ginevra Cusseau and Lucia Pizzocaro) on *P. alternata*. In this last species females were captured from a cluster of nests, individually marked and transported in black containers for various distances and in different directions from the nest, then released. Their presence on the nest was checked the following days. At least one of the females released from a distance of 576 m in a mixed forest and one in open landscape were found again on their nests after some time, but the distance fell to 335 m when the wasps had to travel back entirely through the forest. These rough trials indicate that these wasps have a reasonably good homing performance and wide home range which, in any case, is also influenced by the landscape characteristics. For comparison we can note that in *Polistes dominula* homing performance (percentage of released individuals returning to the nest) in an open landscape is quite high within 400 m, to fall around 50 % at a distance of 2,000 m (Ugolini and Cannicci 1996).

3.5.5 Collection of Nest Material and Nest Construction

Building behaviour and related activities (including ant guard construction) are fully treated in Chap. 6. Material is mainly collected during the foundation phase but almost never in mature colonies where the re-use of old cell material furnishes all the necessary for construction and enlargement of new cells. In *P. nigricans serrei* this activity reaches a maximum in early morning and collecting trips are usually preceded by the collection of water from available sources, such as dew or rain drops on leaves (Turillazzi and Pardi 1982). Very rarely I observed collection of material in the field but old or alien nests can be the source of material in captivity.

3.5.6 Ant Guard Construction

This behaviour is only found in species of the *P. jacobsoni* group, in some species of *Eustenogaster* and only in *L. tutua* in the genus *Liostenogaster*. The behaviour with which the Dufour's gland secretion is applied on the nest substratum in similar in *Eustenogaster* and *Liostenogaster* but fundamentally different in the genus *Parischnogaster* (see Chap. 6). This activity, performed by all the females of a colony, is mainly practised in the afternoon hours, at least in *P. nigricans serrei* (Turillazzi and Pardi 1981).

3.6 Social Behaviour

Social behaviour indicates all those interactions between adults which characterise colonial life. These include both peaceful and aggressive patterns.

3.6.1 Food Exchange

In *P. jacobsoni* (taken as an example for various other species) when foragers return to their nests, females and males run to meet them and try to grab the wads of food. They start chewing the food whilst trying to tear off pieces of bolus, holding their heads against those of the forager, moving their body and head back and forth.

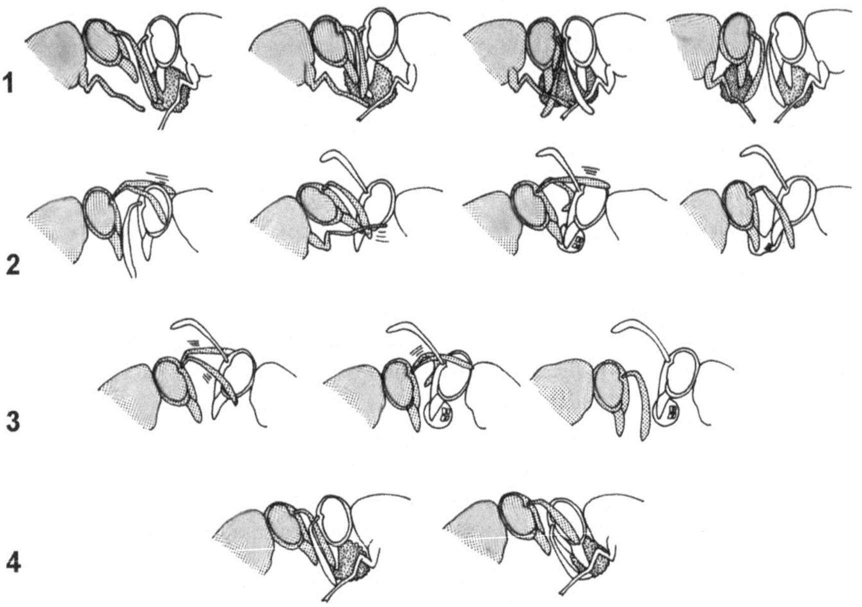

Fig. 3.11 Schematic representation of various kinds of adult–adult interactions; in grey the initiator wasp (1) sharing of solid food, (2) trophallactic exchange of regurgitated liquid (from the white wasp), (3) dominance interaction with the white subordinate female emitting a drop of liquid, (4) the white wasp avoids sharing food with a soliciting female (or male) (from Turillazzi 1990c, re-drawn)

The solicitors will also often use their anterior legs to grasp the boluses, but rarely perform antennal solicitations. The foragers can avoid sharing food, shielding it by lowering their heads toward the thorax. Sometimes females do this when solicited by very aggressive males (Turillazzi 1988; personal observation) (Fig. 3.11).

Fluid requests are just a little different because, usually, the beggar wasps antennate their partners calmly, delivering light blows on the side of their heads, near the antennal sockets, but sometimes they will bite them on the lower part of their clypei and mouth parts and caressing the sides of their head with their anterior legs. The solicited wasps stop with their antennae apart and emit drops of fluid which are sucked up completely, or in part, by the beggar wasps. Fluid requests can also be observed even when the foragers have been back on the nest for some time. Sometimes, for example, wasps that have recently taken food from a larva are solicited to regurgitate by their nestmates. The same sequences and positions are similar in all species where this has been observed (Fig. 3.11).

Trophallactic exchange is the most common interaction observed in *L. vechti* (Turillazzi 1990a) and *L. flavolineata* (Samuel 1987) colonies. I observed 140 of these interactions on 28 nests of the former species in a period of 8 h, which occurred especially on the return of foragers. I observed the tendency to keep the antennae almost motionless and the gastrum with the peduncle gently raised in both the demanding and soliciting wasp. Particularly intense requests may be accompanied by continuous wing flipping (see Videoclips 3.7 and 3.8).

3.6.2 Dominance–Subordinance

Dominant individuals in relatively large colonies usually crawl round their nests quite "sure" of themselves and move towards other individuals they meet on their walk. On such occasions subordinates may try to avoid dominants or immobilise themselves by freezing on the spot. They turn their heads to one side and, after the dominant female has antennated them briefly, emit a drop of fluid between their mandibles that may or may not be sucked up by the dominant. Avoidance performed by subordinate females at the passage of a dominant individual is particularly evident on linear nests of some species (*P. nigricans serrei, P. jacobsoni*) but are also appreciable in other species (Turillazzi and Pardi 1982) (Fig. 3.11) (see Videoclips 3.9 and 3.10).

In *L. vechti* dominance episodes are quite rare but can be recognised from the submissive attitude adopted by the subordinate female when she remains motionless and is bitten on the head or face by the dominant, incumbent individual. The subordinate lowers her head, falls in akinesis and can emit a drop of fluid which the dominant can take or not (Turillazzi 1990a). The whole procedure resembles the dominance–subordinance interactions described for other social wasps, *Polistes* in particular (Pardi 1942).

For *L. flavolineata* Samuel (1987) put the intensity of various dominance interactions in order according to a scale. A subordinate individual can retreat with lowered antennae upon the advancement of a more dominant one which holds her antennae well raised. In some cases the dominant can chase the opponent around the nest but the whole interaction does not include any body contact between the two individuals (this was the most common interaction observed (76 %) out of a total of 872 actions over 102 h of observations on a single nest). Head contact between a dominant and a subordinate can occur when the first touches the head of the second with its antennae and mandibles and the second one lowers its head. Afterwards it can flee or remain motionless under the dominant individual standing over her. In a further escalation of aggression, the dominant can use its mandibles to grasp the legs, or thorax, of the subordinate and flex the gaster attempting to sting the other wasp, with the clear intention of dislodging it from the nest. In some cases the two opponents can grapple together and even fall to the ground, and Samuel reports that in two of these contests the bitten female was severely injured (Samuel 1987).

3.6.3 Fighting

Aggressive behaviours can commonly be observed in large colonies where the density of individuals (males and females) is high, but they are not so common in small colonies. In *L. vechti* the most fierce fights may be preceded by antennal skirmishes of two wasps facing each other, with out-stretched antennae, and end up

Fig. 3.12 Antennal fencing between two females of *L. vechti* (photo by D. Baracchi)

in grasping and falling fights. During antennal fencings, which can last for tens of seconds, the wasps hold their abdomens bent, in the threatening position, and may perform abdominal slaps before grasping the contender and fall from the nest (Fig. 3.12).

The fighters do not fall to the ground but separate in mid-air before returning to the nest to fight again. Such battles can last as long as several hours and can start again the next day, after a break for the night (Samuel 1987; Turillazzi 1990a). I had the opportunity to observe a scene like this on a nest of *L. vechti* when I collected all the population for marking the various individuals. While the nest was unattended, a vagrant female took it over and began fierce combat against the dominant female when she returned to the nest. The contest ended two days afterwards with the acceptation of the alien female on the nest in a subordinate position (Turillazzi 1990a).

Three cases of prolonged antennal fencing were also observed by Samuel (1987) in *L. flavolineata*.

3.6.4 Male Clustering

Males of various species of *Parischnogaster* (Turillazzi and Pardi 1982; Hansell 1986; Turillazzi 1988), *Stenogaster concinna* (Spradbery 1975), *Anischnogaster* (Spradbery 1989; Turillazzi and Hansell 1991) have been observed forming unisexual clusters of several individuals on thread like substrata (Fig. 3.13).

These male clubs originate from the congregation of individuals, coming from different colonies, which have been chased away by the females, and are usually active for various lengths of time. Male mates interact with each other but no particular study has ever been carried out on the social behaviour of these teams. We know, however, that these aggregations can play a relevant role in the foundation of new nests, at least in *Parischnogaster*. In two cases in the wild (Turillazzi and Pardi 1982) and in six cases in captivity (Turillazzi and Francescato 1989), in

Fig. 3.13 Unisexual groups of male *Parischnogaster* sp. (photo by R. Innocenti)

fact, we observed females of *P. mellyi* founding new nests exactly on the substrata, already held by male clubs that they had joined about 2 weeks before. Only those foundresses which afterwards were able to chase all the males away had successful colonies, but it seems probable that females use these male clusters as indicators of suitable nesting places, good for the strength of the substrata and for the relative absence of predators (Turillazzi and Francescato 1989).

References

Baracchi D, Hashim R, Teseo S, Turillazzi S (2009) Basic social biology and nest architecture of *Liostenogaster topographica* Turillazzi 1999 (Hymenoptera Stenogastrinae). Trop Zool 22:15–25

Baracchi D, Dapporto L, Turillazzi S (2011) Relevance of wing morphology in distinguishing and classifying genera and species of Stenogastrinae wasps. Contrib Zool 80:191–199

Beani L (1996) Lek-like courtship in paper wasps: "a prolonged, delicate, and troublesome affair". In: Turillazzi S, West-Eberhard MJ (eds) Natural history and evolution of paper wasps. Oxford University Press, Oxford

Beani L, Turillazzi S (1999) Stripes display in hover wasps (Vespidae-Stenogastrinae): a socially costly status badge. Anim Behav 57:1233–1239

Beani L, Sledge MF, Maiani S, Boscaro F, Landi M, Fortunato A, Turillazzi S (2002) Behavioural and chemical analysis of scent-marking in the lek system of a hover-wasp (Vespidae, Stenogastrinae). Insect Soc 49:275–281

Brown RE, Macdonald DW (1985) Social odours in mammals. Clarendon Press, Oxford

Coster-Longman (1998) Ecological factors in the agglomeration and evolution of sociality in the Stenogastrinae (Hymenoptera Vespidae). PhD thesis, Università di Firenze, Italy
Field J, Foster W, Shreeves G, Sumner S (1998) Ecological constraints on independent nesting in facultatively eusocial hover wasps. Proc R Soc B 265:973–977
Fortunato A, Coster-Longman C (2000) Reproductive behaviour and relative anatomical structures for the production of pheromones and luminous signals in male Metischnogaster (Vespidae: Stenogastrinae). Insect Soc Life 3:27–28
Fortunato A, Turillazzi S (2012) Dufour's gland and massive use of its secretion drove the evolution of sting's morphology and function in hover wasps (Hymenoptera Stenogastrinae). Arthropod Struct Dev 41:259–264
Francescato E, Massolo A, Landi M, Gerace L, Hashim RB, Turillazzi S (2002) Colony membership, division of labour and genetic relatedness among females of colonies of *Eustenogaster fraterna* (Hymenoptera, Vespidae, Stenogastrinae). J Insect Behav 15:153–170
Hansell MH (1982) Brood development in the subsocial wasp *Parischnogaster mellyi* (Saussure) (Stenogastrinae Hymenoptera). Insect Soc 29:3–14
Hansell MH (1983) Social behaviour and colony size in the wasp *Parischnogaster mellyi* (Saussure), Stenogastrinae (Hymenoptera, Vespidae). Proc Konin Neder Akad Wetens Ser C 86:167–178
Hansell MH (1986) Colony biology of the stenogastrine wasp *Holischnogaster gracilipes* (Van der Vecht) on Mount Kinabalu. Entomol Mon Mag 122:31–36
Hansell MH (1987) Elements of eusociality in colonies of *Eustenogaster calyptodoma* (Sakagami & Yoshikawa) (Stenogastrinae, Vespidae). Anim Behav 35:131–141
Iwata K (1967) Report of the fundamental research on the biological control of insect pests in Thailand. II. The report on the bionomics of Aculeate wasps bionomics of subsocial wasps of Stenogastrinae (Hymenoptera, Vespidae). Nat Life SEA 5:259–293
Jeanne RL (1979) A latitudinal gradient in rates of ant predation. Ecology 60:1211–1224
Landi M, Coster-Longman C, Turillazzi S (1998) Alarm response in a laboratory population of *Parischnogaster mel*lyi (Stenogastrinae). Insect Soc Life 2:59–64
Pagden HT (1958) Some Malayan social wasps. Malayan Nat J 12:131–148
Pagden HT (1962) More about *Stenogaster*. Malayan Nat J 16:95–102
Pardi L (1942) Ricerche sui Polistini. V. La poliginia iniziale in *Polistes gallicus* (L.). Boll Ist Entom Univ Bologna 14:1–106
Sakagami SF, Yamane S (1983) Behavior inventory of *Parischnogaster mellyi*. In: Ohgushi R (ed) Ecological study on social insects in Central Sumatra with special reference to wasps and bees. Sumatra Nature Study (Entomology), Kanazawa University, Japan
Sakagami SF, Yamane S (1990) A behavior inventory of the females of two stenogastrine wasps *Parischnogaster mellyi* and *Liostenogaster vechti* (Hymenoptera, Vespidae). In: Sakagami SF, Ohgushi R, Roubik DW (eds) Natural history of social wasps and bees in equatorial Sumatra. Hokkaido University Press, Sapporo
Samuel C (1987) Factors affecting colony size in the stenogastrine wasp *Liostenogaster flavolineata*. PhD thesis, University of Malaya, Kuala Lumpur
Spradbery JP (1975) The biology of *Stenogaster concinna* Van der Vecht with comments on the phylogeny of Stenogastrinae (Hymenoptera: Vespidae). J Aust Entomol Soc 14:309–318
Spradbery JP (1989) The nesting of Anischnogaster iridipennis (Smith) (Hymenoptera: Vespidae) in New Guinea. J Aust Entomol Soc 28:225–228
Starr CK (1984) Rambling remarks on Philippine wasp-mimics. Sphecos 9:11–13
Turillazzi S (1983a) Patrolling behaviour in males of *P. nigricans* and *P. mellyi*. Rendiconti Acc Naz Lincei Cl Scienze MFN 72:153–157
Turillazzi S (1983b) Extranidal behaviour of *Parischnogaster nigricans serrei*. Z Tierpsychol 63:27–36
Turillazzi S (1985a) Function and characteristics of the abdominal substance secreted by wasps of the genus *Parischnogaster*. Monit Zool Ital 19:91–99

Turillazzi S (1985b) Egg deposition in the genus *Parischnogaster* (Hymenoptera Stenogastrinae). J Kansas Ent Soc 58:749–752

Turillazzi S (1985c) Brood rearing behaviour and larval development in *Parischnogaster nigricans serrei*. Insect Soc 32:117–127

Turillazzi S (1988) Social biology of *Parischnogaster jacobsoni* (Du Buysson) (Hymenoptera, Stenogastrinae). Insect Soc 35:133–143

Turillazzi S (1990a) Social biology of *Liostenogaster vechti* Turillazzi 1988 (Hymenoptera Stenogastrinae). Trop Zool 3:68–87

Turillazzi S (1990b) Notes on the biology, social behaviour and nest architecture of *Metischnogaster drewseni* (Saussure) (Hymenoptera Stenogastrinae). Boll Zool 57:331–339

Turillazzi S (1990c) Socialità ed architettura del nido nelle vespe Stenogastrinae (Hymenoptera, Vespidae). Atti Acc Naz Ital Entomol Rendiconti XXXVIII:1–24

Turillazzi S, Francescato E (1989) Observation on the behaviour of male stenogastrine wasps (Hymenoptera, Vespidae, Stenogastrinae). Actes Coll Insectes Soc 5:181–187

Turillazzi S, Francescato E (1990) Patrolling behaviour and related secretory structures in the males of some stenogastrine wasps (Hymenoptera Vespidae). Insect Soc 37:146–157

Turillazzi S, Hansell MH (1991) Biology and social behaviour of three species of *Anischnogaster* (Vespidae, Stenogastrinae) in Papua New Guinea. Insect Soc 38:423–437

Turillazzi S, Pardi L (1981) Ant guard in the nest of *Parischnogaster nigricans serrei* (du Buysson). Monit Zool Ital 15:1–7

Turillazzi S, Pardi L (1982) Social behavior of *Parischnogaster nigricans serrei* (Du Buysson) (Hymenoptera Stenogastrinae) in Java. Ann Entomol Soc Am 75:657–664

Turillazzi S, Cervo R, Dani FR (1997) Intra- and inter-specific relationships in a cluster of Stenogastrine wasp colonies (Hymenoptera; Vespidae). Ethol Ecol Evol 9:385–395

Ugolini A, Cannicci S (1996) Homing in paper-wasps. In: Turillazzi S, West-Eberhard MJ (eds) Natural history and the evolution of paper wasps. Oxford University Press, Oxford

Williams FX (1919) Philippine wasp studies, part 2: descriptions of new species and life history studies. Bull Exp Sta Hawaii Sugar Planters' Assoc Entomol 14:19–186

Chapter 4
Colonial Dynamics

4.1 Factors Influencing Colony Dynamics

The existence, consistency and characteristics of animal groups are dictated by innumerable factors, external and internal to the groups themselves. Environmental ones, such as climate and predators, can be decisive in promoting living together if this confers advantages to the individuals; while characters intrinsic to the biology of a certain species, such as parental care and genetic structure of populations, may represent facilitating factors which can mould the social organisation of the groups.

Groups are, in any case, initiated by single individuals and individual characteristics and predispositions may have important consequences in the structure of the consortium to which they belong.

Hover wasps are social animals but, as we have seen, they have some peculiarities which differentiate them from other social wasps and are also important in making their colonies diverse from other animal societies. Ever since the first studies on these wasps, various authors realised the peculiarities of their social behaviour and considered them as a crucial group for understanding the evolution of insect sociality (Williams 1928; van der Vecht 1977; West-Eberhard 1978). In this chapter, I shall analyse hover wasps as social animals trying to figure out, on the basis of the information available at present but which has greatly increased over the last 30 years, the main characteristics of their colonies and to answer the question about their importance in a generalised theory on social evolution.

I shall start with an analysis of the environmental characteristics of the sites where various species live and of the principal a-biotic and biotic factors that regulate their social life, including the adaptations characteristic of the species and the possible different life histories of individuals. Then I shall pass to illustrating the colonial development of some species and the main features of their social structure with the onset of division of labour between individuals. I shall end by examining the characteristics of the mechanisms that regulate their social organisation.

S. Turillazzi, *The Biology of Hover Wasps*, DOI 10.1007/978-3-642-32680-6_4,

4.1.1 The Environment

The geographical distribution of Stenogastrinae is limited to the East Asian tropics (see Chap. 1). Hover wasps are rainforest inhabitants; most species live in quite humid micro-environments, preferring to nest under the banks along the forest streams or near waterfalls, or on the ceiling of caves hidden in the jungle. Probably, many species are still unknown to science precisely because of the characteristics of their habitat and undoubtedly most of the better known species are those which frequent human settlements and build their nests on substrata represented by buildings or other human artefacts.

4.1.1.1 Climate Factors

Temperature

In the tropics, temperature is not usually a serious limiting factor to the active life of insects. Indeed, in lowland and hill forest environments there is no particularly adverse season during which colonies of hover wasps are absent as, on the contrary, occurs for species of social wasps in temperate regions. Figure 4.1a, for example, gives the average temperature recorded at the Gombak Field Centre of the University of Malaya (Central Peninsula Malaysia, 260 m asl).

Some species, however, live exclusively above a certain altitude, where the temperature is almost constant over the course of the year (Fig. 4.1b) but may have minor daily variations. Bukit Fraser (Fraser Hill) is a pleasant tourist locality on the Central Mountain Range of Peninsular Malaysia which is situated above 1,000 m asl and reaches an altitude of 1,600 m at its highest point. Here several researchers, starting with Yoshikawa and his colleagues, took turns over the year in collecting samples and studying populations of hover wasps. Also my group, in this place, performed various studies on a beautiful species of *Eustenogaster*, *E. fraterna*, which lives above 800–900 m altitude (Turillazzi and Gerace 1992; Francescato et al. 2002; Landi et al. 2003). This species presents some interesting characteristics of social behaviour which differ from those of congeneric species living at lower altitudes, such as *E. calyptodoma* (Hansell 1987).

Variations in temperature can be even greater (Fig. 4.2a, b) for species which inhabit the northern limit of the distribution range. *Eustenogaster nigra* is a species which lives in Northern Vietnam and in the Southern China State of Yunnan, as far as the Hong Kong region (Saito et al. 2006, 2009; Barthélémy 2008, 2009). Recently, some Japanese and Vietnamese entomologists discovered that this species presents a winter diapause. Dormant males and females gather together to form hibernating clusters inside nests without any immature brood, until the next good season when mating also takes place (Saito et al. 2006, 2009).

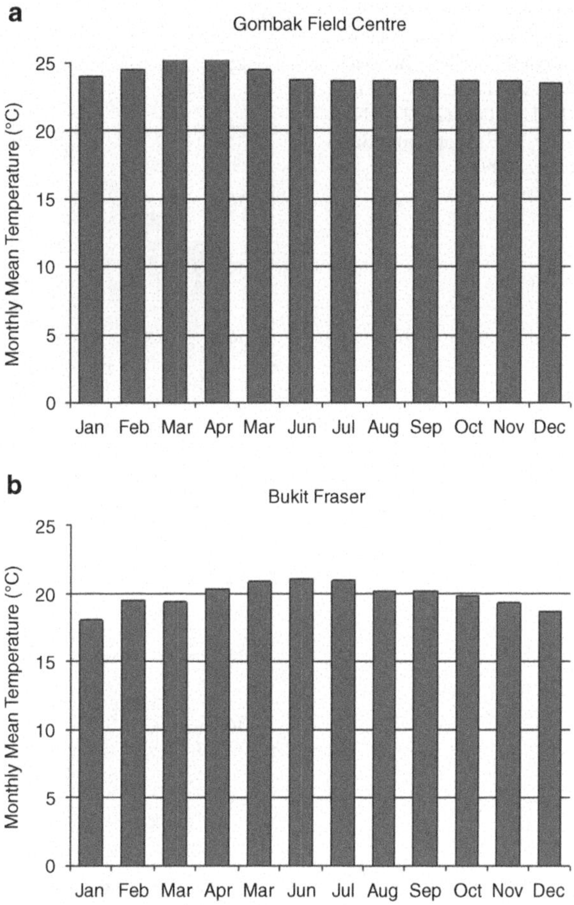

Fig. 4.1 (**a**) Annual variation in temperature recorded at Gombak Field Centre of the University of Malaya (3° 19′N, 101° 45′E, Central Peninsular Malaysia, 260 m asl) in 1967. (**b**) Annual variation in temperature recorded at Bukit Fraser (3° 42′ N, 101° 44′E, Central Peninsular Malaysia, 1,300 m asl)

Rainfall

Even if tropical climates do not experience significant changes in temperature over the year, the southeast Asia tropics have a wet and dry season caused by the seasonal shift in the monsoons. Figure 4.3 gives the average rainfall, over a period from 1978 to 2007, recorded at Genting Sempah, a locality on the central Mountain Range of Peninsular Malaysia where I performed much of my research.

The main peak for rainfall occurs in November while a further lower peak generally occurs in May. Minimum rainfall is usually observed in January–February.

Charlotte Samuel was the first to connect data on rainfall periods with the population biology of *Liostenogaster flavolineata*, a hover wasp that she studied in various localities in Peninsular Malaysia. In her PhD thesis she pointed out how varying intensities of rainfall were important for variation in the adult and immature brood population of this species. In particular, the maximum total female and male populations were observed at times of minimum rainfall (i.e. in the driest period of

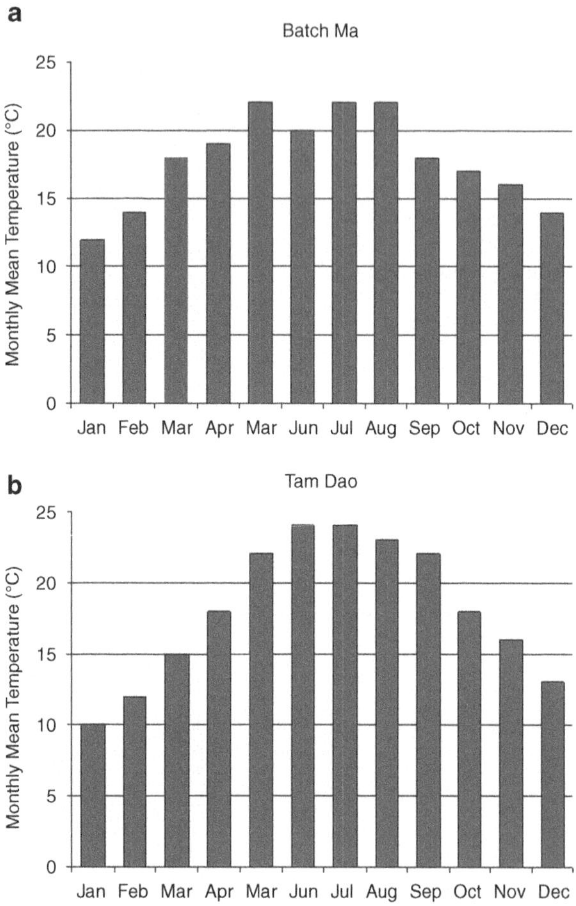

Fig. 4.2 Annual variation in temperature at two localities in North Vietnam (**a**) Bach Ma (16° 12′N, 107° 53′E) (**b**) Tam Dao (21°27, 5N, 105° 38′E) (data from Saito et al. 2009)

the year), while the minimum adult population coincided with times of maximum rainfall. An unusually long dry season in 1983 coincided with a slower increase in the wasp population at the end of the season itself (Samuel 1987).

Recently Cronin et al. (2010) analysed data accumulated over various research campaigns in some of the nesting sites of the same species in Malaysia, reaching further important conclusions. In particular they observed, again with marked differences in the rainfall periods of the single annual cycles, several and temporally distinct peaks of brood production. The number of newly-emerged females and mature brood were positively correlated with temperature and negatively associated with rainfall. Peaks of brood production also coincided with the emergence of larger individuals. The authors stress the fact that colonies (at least in the observation period) exhibit increased brood production during the warmest part of the year and between two wetter periods (suggesting a substantial convergence with the observations of Samuel 1987), and that colony demography in *L. flavolineata*

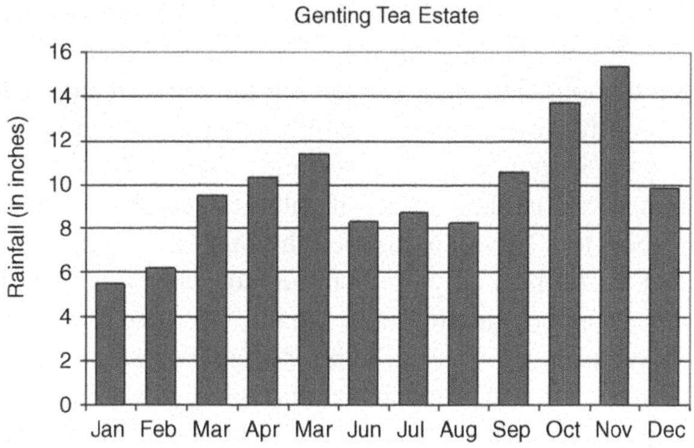

Fig. 4.3 Average monthly rainfall recorded at Genting Sempah (3° 21′ N, 101° 47′ E; Central range of Peninsular Malaysia, 650 m asl) during the period 1978–2007

appears to be regulated by seasonal variations and could be, at least in part, environmentally driven (Cronin et al. 2010).

In some cases, heavy rains can be a limiting factor for the survivorship of the colonies. Floods and earth slides can destroy nests and immature brood limiting, "de facto", the overall population for the following months. For example, I have witnessed several cases of clusters formed of hundreds of colonies of *Parischnogaster alternata* and *Liostenogaster flavolineata* under small bridges completely swept away by streams swollen after sudden thunderstorms.

Soil type and the geology in general of a given site can influence the distribution of a species, especially with regard to the availability of places where to nest and the particular materials that can be used for nest construction. The amount of water available in the environment may also be a limiting factor for another reason, as it can condition the quantity of material the wasps can collect for nest construction if they use mud.

4.1.1.2 Biotic Factors

Biotic factors can be predation, disease and prey availability, but also competition, with conspecifics or with species with similar attitudes and characteristics, for resources such as food, water, nest site and mates.

Predators

Like other social wasps, colonies of hover wasps, even if limited in size, are agglomerations "packed" with proteins which can be appetising morsels for various small predators. The main predators of these wasps, in any case, are reported to be hornets of the genus *Vespa*, which also limit the populations of other social wasps in South East Asia (Matsuura and Yamane 1990) (Fig. 6.59). In some cases the impact of these predators on a local population can be exceptionally heavy. In a recent survey of colonies of *Liostenogaster vechti* in a location near Fraser Hill, the Jeriau waterfall, we realised that almost all of the hundreds of colonies forming dozens of clusters on buildings in the recreation area, had very young larvae with a steady, adult population but no pupae nor fully grown larvae. In the end we discovered the reason for this unusual situation when we found various hornets of the species *V. tropica* systematically harvesting colonies in some exposed clusters. The strange thing was that the number of colonies and clusters had undoubtedly increased with respect to a survey performed the previous year; one explanation for this could be that the hornets only prey on the more developed immature wasp brood and leave the younger, smaller ones intact. As predation is carried out only by some individuals, which seem to be specialised in this kind of task, and does not involve recruitment, it is probable that the colonies have in any case enough time to flourish in periods when such individual hornets are not active, or hornet colonies are not established in the surroundings.

V. tropica is also the main predator of *P. nigricans serrei* colonies in West Java where it destroys the cells to remove larvae and pupae, leaving those with eggs and little larvae intact. It is not rare for the same colonies to suffer consecutive raids when the pupae and larvae have developed again. In seven predation raids on seven colonies, I observed the hornets removing over 90 % of the big larvae, 75 % of the pupae and only 6.7 % of the eggs and little larvae (Turillazzi 1985a).

Mike Hansell also observed *V. tropica* preying on colonies of *Parischnogaster mellyi* in Thailand; raids by this hornet can also be detected from the characteristic damage they cause, consisting in holes torn out on the sides of the cells (Hansell 1982). Jeremy Field (personal comm.) has actually observed *V. tropica* carrying away an entire nest of *Parischnogaster alternata*!

Other species of hornets are involved in hover wasp predation: Williams (1928), for example, mentions various consecutive attacks on a nest of *Parischnogaster depressigaster* by the hornet *Vespa deusta* in the Philippines.

As reported by Jeanne (1979), ants are the most important predators of social wasps in tropical environments and this can also be true for hover wasps. Actually, these wasps are careful to nest in places where ants are very infrequent but, in any case, it is not unusual to find nests plundered by various species of these efficient predators. Samuel, for example, observed workers of *Technomyrmex* sp. raiding colonies of *L. flavolineata* (Samuel 1987).

Spiders can catch hover wasps as they try to pluck small prey from their webs and I have observed such casualties twice; or else flying wasps can become accidentally caught in a spider web during their flights. Other casual predators can be included in the list: for example, once I observed a forest nematode attacking a cluster of males of *Parischnogaster* on a rootlet and entrapping an individual which was then killed and sucked while the others remained perfectly still in the cluster without moving (Fig. 4.4) (Turillazzi and Pardi 1982).

Vertebrate predators have rarely been observed: geckoes usually prey on colonies of species which are more tied to human environments, while Starr (pers. communication) in the Philippines observed bats preying on nests of a species of *Parischnogaster* suspended from the vault of a cave.

Fig. 4.4 A nematode attacking a cluster of *Parischnogaster* males

Parasites

Parasitoids such as Tachinid flies are quite common in some stenogastrine species. Spradbery, for example, found cocoons of an undetermined tachinid in 4 out of 34 nests of *Stenogaster concinna*, some of which were themselves hyperparasitised by eulophid flies (Spradbery 1975). The same author also found 9 out of 20 nests of *Anischnogaster iridipennis* attacked by the tachinid *Euvespivora decipiens,* with a rate approaching 80 % parasitism in individual nests (some hyperparasitized by the eulophid of the species *Nesolynx*) (Spradbery 1989). In other species of *Anischnogaster* Hansell and I (Turillazzi and Hansell 1991) found large numbers of cells occupied by tachinid puparia while one more species of these parasitic flies (*Petrorossia* sp.) has also been reported by Samuel (1987) as a parasitoid of *Liostenogaster flavolineata*. In this last species, *Melittobia* (Chalcidoidea) is also considered as the chief parasite of the immature brood, together with *Anthrax* sp. (Bombyliidae) (Samuel 1987). Also Krombein (1991) reports about seven eulophid larvae (prob. *Melittobia*) found in a larva of *Eustenogaster eximia*.

Ichneumonid parasitoids of the genus *Theronia* have been found by Williams (1928) in nests of hover wasps, whilst another eulophid (*Syntomosphyrum* sp.) has been reported as a parasite of stenogastrine larvae by Iwata (1967).

Mites and Psocoptera are usual inhabitants of active and abandoned hover wasp nests (personal observation) and I suspect that adults spend much of their time keeping their nests free from these potential risks to their immature brood. Spradbery (1975), on the other hand, found 53 phoretic mites attached to the wings of a female of *Stenogaster concinna*.

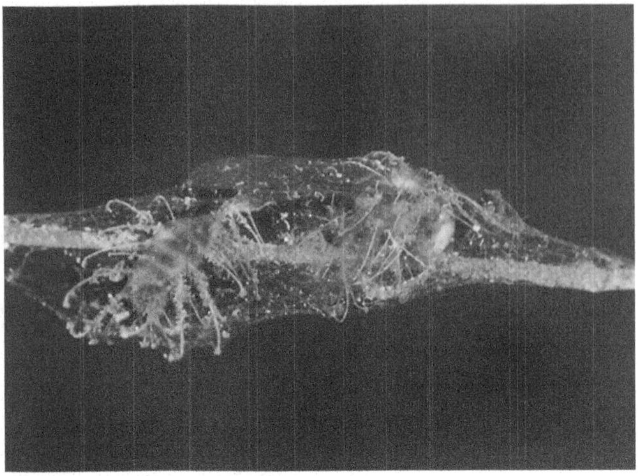

Fig. 4.5 A corpse of a *Parischnogaster* sp. with sporocarps of a pathogenic fungus springing from it

Even though nest usurpation is quite common in various species of hover wasps, obligate socially parasitic species have never been observed. One possible exception could be the long sting females of *Anischnogaster laticeps* that Mike Hansell and I found in a population of colonies of the same species (see Sect. 2.1.3.1).

Pathogens

One of the most important disadvantages of social life in general is the increased possibility of transmission of pathogens and parasites between the adult and immature individuals in a colony. Pathogens, however, are practically unknown in these wasps, evidently because of the lack of information as not one single study has ever been performed on the matter. Pathogenic fungi have been observed on killed individuals. Spradbery, for example, writes about a male *S. concinna* killed by a *Cordyceps* parasitic fungus (Spradbery 1975). I found various fungi, probably of the same genus, on dead male and female individuals of *P. jacobsoni* and *P. mellyi*. The corpses of the wasps were found with all their legs grasping to a thread-like substrata close to conspecific nests. The fungal hyphae enveloped the body and the sporocarps of the fungus sprang out from the corpse (see Fig. 4.5).

Prey

Prey availability can influence the distribution of a species in a given area; however, our information on the prey of hover wasps is rather scanty and limited only to fragmentary news in the literature. Various authors have observed the peculiar habit of species of the genera *Parischnogaster*, *Metischnogaster*, *Anischnogaster* and *Eustenogaster* to pluck small remains of prey from spider webs, but it is still unknown if they prefer any particular items. Examination of prey loads brought to the nest by foragers or found in the coiled mature larvae were performed on *S. concinna* by Spradbery (1975) who found fragments of bodies of small Diptera, probably cecidomyiid midges. Small spiders brought to the nest by *P. mellyi* were observed by Sakagami and Yamane (1990). Samuel (1987) found a high proportion of pieces of ants' bodies (mainly male alates of probably *Camponotus* and *Polyrachis* and formicine species) and termites (male and female alates) in samples collected from mature larvae of *L. flavolineata*. These last data are all we have on a species of *Liostenogaster,* a genus which, owing to its less manoeuvred flight, has never been seen foraging at spider webs, probably because it is unable to do so. If these wasps do prey only on swarms of alates of ants and termites, such finds would be rather unpredictable and real limiting factors for the distribution and consistency of colonies.

Krombein (1991) performed accurate analyses on the gut contents of larvae of *E. eximia* and on boluses of larval food finding several parts of spiders, ant workers, ichneumonids and braconids; thus it seems that species of the genus *Eustenogaster* prey on a wide range of small arthropods.

I have successfully reared colonies of species of *Parischnogaster* and *Eustenogaster* in captivity. *P. mellyi* proved to be the most easy species to rear, as the females forage on a wide range of small insects (randomly collected with a net in the garden of our Department). They seem to particularly appreciate fruit flies but do not hesitate to take pieces of larger flies, small crickets and termites attached to the glasses of the rearing cages with tiny drops of water.

4.2 Colony Development

4.2.1 Colony Size

Social interactions in animals are more easily studied if the groups are composed of a limited number of individuals. Colonies of hover wasps are good subjects to study in this respect as the individuals which one can find in a colony can usually be counted on the fingers of one hand. This permits careful observation of the behaviour of single individuals and of social interactions. The maximum number of individuals observed in colonies of various species is reported in Table 4.1.

In some species the average number of females per colony is less than two: this seems to be a contradiction of the term "social insects"! Actually some species can be defined as "eusocial" only for some brief periods of their colony cycle. In Papua New Guinea, studying hover wasps of the genus *Anischnogaster* together with Mike Hansell, we at first were able to find only nests with solitary adult females. For some time we imagined that we had found the very first solitary hover wasp, which

Table 4.1 Maximum number of cells and maximum number of females per nest reported for various species of stenogastrine wasps with indication of the literature source

Species	Max cells	Max females	Max males	Max adults	Max brood	References
Anischnogaster sp.A	20	2	0	2	9	Turillazzi and Hansell (1991)
Anischnogaster sp.B	15	1	0	1	7	Turillazzi and Hansell (1991)
A. iridipennis	13	2	1	3	8	Spradbery (1989)
A. laticeps	8	2	0	2	6	Turillazzi and Hansell (1991)
Eustenogaster calyptodoma	13	4	1	5	11	Hansell (1987)
E. eximia	25	2	2	4	18	Krombein (1991)
E. fraterna	19	6	5	11	18	Turillazzi and Gerace (1992)
E. hauxwellii	12	3	1	4	10	Turillazzi (personal observation)
E. micans	15	3	1	3	10	Turillazzi (personal observation)
E. nigra (summer nests)	17	1	0	1	16	Saito et al. (2009)
E. scitula (summer nests)	16	6	2	8	12	Saito et al. (2009)
Liostenogaster flavolineata	110 (123)	10	12	11	55	Turillazzi (personal Observation); Samuel (1987); Ohgushi et al. (1990)
L. nitidipennis	53	6	?	20	?	Turillazzi (personal observation)
L. pardii	53	6	1	7	?	Turillazzi and Carfi (1996)
L. topografica	83 (113)	8	7	38	48	Baracchi et al. (2009), personal observation
L. vechti	31	7	4	9	20	Turillazzi (1990a)
Metischnogaster drewseni	17	2	1	3	8	Turillazzi (personal observation)
Parischnogaster alternata	40	13	15	10	28	Ohgushi et al. (1990); Turillazzi (1986)
P. gracilipes	54	8	2	9	32	Hansell (1986)
P. jacobsoni	48	6	6	10	33	Turillazzi (1988)
P. mellyi	51 (100)	9	4	9	46	Fanelli et al. (2005); Hansell (1981); Ohgushi et al. (1990)
P. nigricans	42	11	8	16	30	Turillazzi (1985a)
P. striatula	25	5	5	9	15	Bongiovanni (1998), Turillazzi (personal observation)
Stenogaster concinna	17	2	0	2	11	Spradbery (1975)

Data derive from different authors. Extraordinary cell numbers for colonies of *L. flavolineata* and *P. mellyi,* reported by Ohgushi et al. 1990, and for *L. topografica,* reported by Baracchi et al. 2009, are given in brackets

various researchers had indeed foreseen and stressed the relatively primitive social organisation of this group. However, after some days of careful observation, we realised that some of the nests were also inhabited by a less faithful female which was present only for brief periods. An anatomical examination showed that these females were young and not yet fertilised and had undeveloped ovaries. We then ascertained that the daughters of the foundress female remained with their mother for a while to help her in foraging and raising the larvae but left the colony when they were probably able to found a nest on their own (Turillazzi and Hansell 1991).

Colonies composed of only two females, even for short periods, are also found in other genera: *Metischnogaster* is one of these. The nest of these wasps is quite difficult to find because it is very well camouflaged. It looks like a dry stick enmeshed in the long hairy hyphae of *Marasmius*, a fungus which attacks dead forest trees. The nests of this genus usually have no more than a dozen or so cells and the largest colony I found was composed of only three individuals: two females of different age and one male (Turillazzi 1990b). Such small colonies are practically invisible in the forest and their small size can be considered as an adaptation to pressure exerted by vertebrate and invertebrate predators: an evolutionary choice! (see Fig. 6.56).

In any case, the size of the colonies of other species is never very large. The maximum number of cells I have ever counted in an active nest was 110 in *Liostenogaster flavolineata*, while other species can occasionally reach a hundred cells. The number of adults stays below ten and rarely goes beyond this number, even counting the individuals of the two sexes. On a nest of *Liostenogaster topographica* we once found 32 individuals of both sexes: this is probably the largest population ever reported for a colony of Stenogastrinae (Baracchi et al. 2009).

4.2.2 Nest Foundation

The foundation of new colonies takes place all year round in tropical climates. In a survey performed over 8 months (from November 1981 to July 1982) Samuel did not find any significant difference between the average number of foundations per week, recorded in clusters of *L. flavolineata* nests, for the wet and dry seasons; however, according to the data furnished by the same author, it actually seems that foundation of new nests is practised more in dry periods than wet (Samuel 1987, p. 290). Field et al. (1998) put these data in relation with the highest density of females which occurs in the dry season.

Colony foundation in social wasps can have various modalities: usually nests are initiated by single foundresses (haplometrosis or solitary foundation). These females can be joined quite early on by other, potentially equal, females which contribute to the construction of the nest (in this case we speak of associative foundation or pleometrosis). When we find part of a colony leaving an established one and founding a new nest, we refer to it as a swarming foundation: a swarm is usually formed of reproductive and non-reproductive individuals and can be found in those societies which present a great number of individuals. In the Vespidae this mainly occurs in species of the tribes Epiponini and Ropalidiini of the subfamily Polistinae.

In hover wasps we have fundamentally the first two modalities of nest foundation. Actually, solitary nest foundation is the rule in most species while the second has been mainly reported only for species which nest in aggregates. Therefore, a colony begins with the decision by a single individual that implants the very first cell of a nest at a site and in a position that it "retains" the best for the initial investment on its future fitness. Optimal sites for founding a new nest may not be all that frequent, even in the jungle, and founders must be driven by basic rules which determine the choice of the best place available. Once I performed a very simple experiment by placing some electric wires, 30 cm long, hanging from the ceiling of the open basement of the main building at the Gombak Field Centre (Malaysia), a site where some colonies of *Parischnogaster* were already present. After only one week I observed that all the wires had already been used as substrata for new nests by females of *P. mellyi*, *P. jacobsoni* and *P. striatula*. It seems incredible how, even in a jungle (with billions of thread-like substrata of all kinds), wasps can suddenly invest their time and energy on an artificial wire. This was probably due to the protected position from weather inclemency under the shelter of the human building.

Sometimes females seem to have a clear idea about the position where they prefer to implant their nest, but this is not always possible and they have to face opponents which already occupy a good place. This is what seems to happen in the clusters of *L. vechti* and *P. alternata* colonies where the central positions are particularly desirable (Coster-Longman 1998; Coster-Longman et al. 2002) and in *L. flavolineata* where positions in darker parts of the clusters are preferred (Field 2008). In these conditions the construction of new nests usually takes place at the borders of the cluster, unless a vacant position becomes available due to the destruction of an old nest for various reasons. In some cases, dozens of nests must be rebuilt from the start and in this case we can speak of a particular type of associative foundation in which all the adult females belonging to a cluster of colonies must reconstruct their nests. This is not a rare occasion; in fact, entire clusters of nests can be destroyed by natural events (floods, land-slides) or by man himself. In some of these cases we saw that the adult wasps usually re-build their nests in exactly the same position, as occurred in a small cluster of *L. vechti* the nests of which had been experimentally moved (Coster-Longman 1998).

We observed a special type of solitary foundation in *P. mellyi* and *P. nigricans serrei*. In this case foundresses joined male clubs on particular substrata which was used day after day. Probably the females build their nests there taking advantage of the fact that the males had already experimented for themselves the good qualities and predator-free position of the substrata (see Turillazzi and Francescato 1989).

4.2.2.1 Associative Foundation

Associative foundation has been reported in *P. alternata* (Turillazzi 1985b), *L. flavolineata* (Samuel 1987) and observed in *L. vechti* (personal observations) while it is occasional in *P. jacobsoni* (Turillazzi 1988) and in *P. striatula* (personal observation) (Fig. 4.6). All these species, except for the last two, usually nest in clusters, forming great aggregations of colonies. I think that two things may favour the aggregation of females in foundation: the high density of individuals at the site and the particular architecture of the nest which does not keep the colony sufficiently safe from terrestrial predators such as ants, small cockroaches or spiders. In *P. alternata*, for example, the continuous presence of at least one female may assure active defence against predators but, above all, against conspecific females

Fig. 4.6 Associative foundation in *Parischnogaster striatula* (photo by D. Baracchi)

coming from neighbouring colonies. Such females are usually busy stealing material and even Dufour's gland secretion and eggs from other nests (Turillazzi 1985b; Turillazzi et al. 1997).

A similar consideration can be made for the associative foundations observed by Samuel (1987) who, however, found only a minority of two-three females foundations compared with single female foundations (4 vs. 38) in a cluster of *L. flavolineata* colonies. In *L. vechti* I found 7 two-female foundations (presumed as such from the small number and size of the cells) in a big cluster at Fraser Hill: all the associations but one were formed by one fertilised female with large ovaries and one un-fertilised female with undeveloped ovaries (personal observations).

4.2.3 Colony Cycle

These wasps are incredibly lazy! I remember staying for hours and hours observing one colony (which particular species does not matter!) without noting the slightest movement or the minimum interaction between the various individuals. Then, all of a sudden, a female arrives from outside, with a ball of food or a sip of nectar; at this point all the wasps wake up, force the newly arrived companion to drop her load, split it up among themselves and begin to patrol the nest briefly. In just one minute all activity ceases, each wasp takes up its old position and calm reigns once more. From this description one can realise how the development of a society of hover wasps can be an amazingly long story. The egg-laying capacity of a female is extremely limited if compared, for example, with that of the queen of a colony of *Polistes*. A female of *Parischnogaster* is exceptional, for example, if she lays a single egg in a day. During a study of the Indonesian species *Parischnogaster nigricans serrei* I had to wait days before observing the deposition of a single egg.

4.2.3.1 Colonial Phases

Long term periods of observation on colonies of hover wasps have been performed on only four species: *P. mellyi* (Hansell 1981, 1982, 1983), *P. nigricans serrei* (Turillazzi 1985a; 1989), *L. flavolineata* (Samuel 1987; Field et al. 1998) and *E. calyptodoma* (Hansell 1987).

Colony failures can be high. Over almost one hundred days of observation failure was recorded most frequently on foundations (75 %) in *P. nigricans serrei*. Causes of failure were mainly usurpation by other females, destruction of the nest after various accidents and, above all, abandonment by the nest inhabitants (which was the most frequent cause of breakdown—almost 50%—in foundations) (Turillazzi 1985a). Similarly, the possibility of failure of single-female nests is quite high in *L. flavolineata* (Field et al. 1998). Nests are costly to build and only 10–30 % of the females can expect to survive the very long period of development of immature brood (over 100 days on average in this species) which represents an additional critical constraint on independent nesting (Samuel 1987; Field et al. 1998).

In my study on the social biology of *Parischnogaster nigricans serrei* in West Java I found it useful to distinguish five stages in the colonial cycle. The pre-emergence period (i.e. the period in which none of the immature brood reared in the nest has yet emerged as an adult) can be divided into two stages: foundation (F) (from the start of the first cell to the pupation of the first immature brood) and initial nest (I) (until the emergence of the first adult). In the post-emergence period we can recognise three stages: the first can be called young colony (Y) and can be synthetically defined as "the period which goes from the first imago emergence to the emergence of the first male". The other two phases, that I called Middle-age colony (M) and Mature colony (Ma) have no clear cut between them and, obviously, can depend on the species.[1]

In *P. nigricans serrei* I could reconstruct the colony cycle placing end to end the records from three colonies followed in West Java. The full cycle was estimated

---

[1] For *P. nigricans serrei* I used some absolute and relative markers. If the previous history of a colony is known, the passage from an M to Ma colony can be established as the moment in which the female/male ratio of the individuals emerged on that colony becomes equal or less than 1. If we do not know the previous history of a colony we can use parameters observed in colonies of known history and consider it "mature" in the case where fewer females than males are present on the nest, or the nest has more than 24 cells, or the total number of adults present are more than ten. Mature colony stage can last indefinitely and possibly revert to a previous stage in case of accidents (Turillazzi 1985a).

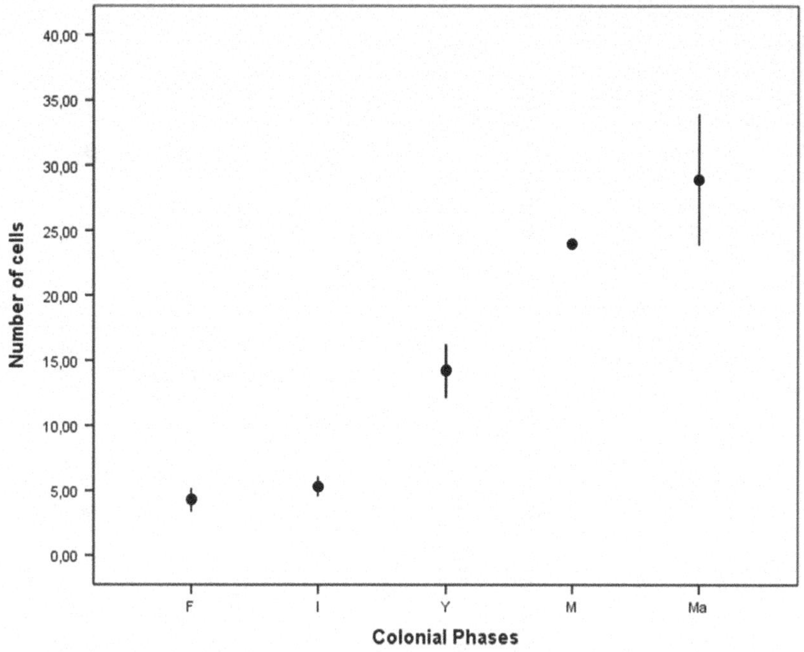

Fig. 4.7 Maximum cell number in colonies of *Parischnogaster nigricans serrei* at various stage of development. For M colonies 24 cells is taken for definition. *Bars* indicate 95 % confidence limits of the average

to last around 7–8 months from foundation to clear signs of senility represented, above all, by the reduction of the number of cells in the nest. Recovery of the nest population can occur, however, at any moment and this can be valid for all the species of Stenogastrinae as nests are often usurped or adopted by wasps coming from outside.

Maximum cell number of colonies in various stages of development (7 Foundations, 9 Initial and 5 Y colonies, 1 M colony and 10 Ma colonies) are reported in Fig. 4.7. We can observe that the increase in cell number is almost constant for colonies in Y and Ma stage and begins to slow down in more mature colonies. Adult female and male emergence and disappearance rates in the colonies, calculated on the total number of adults (emerged or disappeared) in each stage divided by the total number of observation days, is given in Fig. 4.8a, b for colonies at various (Y, M and Ma) stages of development. Total emergences increase from Y to M and Ma colonies but the total expansion rate is maximum in the Y and then begins to decrease in M reaching a minimum in Ma colonies (Turillazzi 1985a).

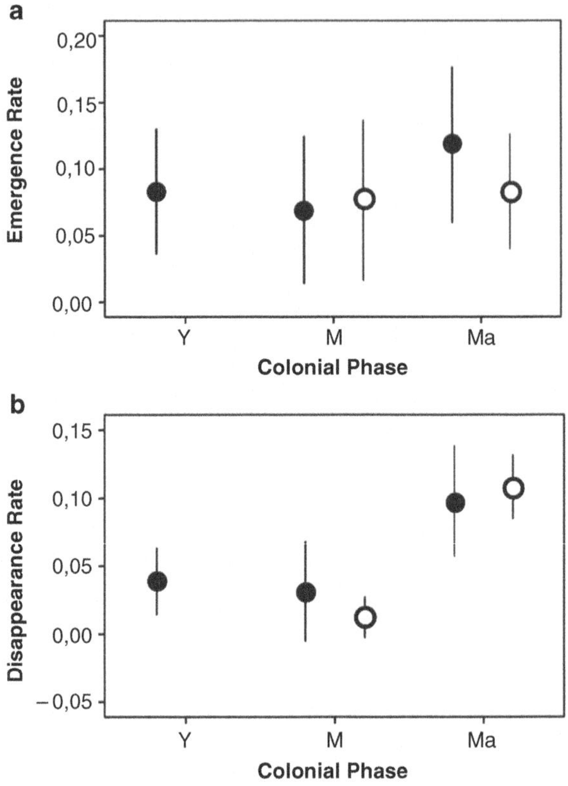

Fig. 4.8 Average emergence (**a**) and disappearance (**b**) rate (individuals/observation days) of adult female (black circles) and male (white circles) individuals in colonies of *Parischnogaster nigricans serrei* at various stage of development. *Y* young colonies, $N = 7$; *M* middle age colonies, $N = 6$; and Ma = mature colonies, $N = 7$). Bars indicate the 95 % confidence limits of the average

4.2.3.2 Sex Ratio

First of all let's give a look at the consistency of the two sexes in the colonies of various species. It is true that, like other social Hymenoptera, the appearance on the nest of the two sexes is slightly unbalanced towards females. In *Parischnogaster nigricans serrei,* that I studied in Java, I found that males begin to emerge after three to five females in new colonies, but then they can abandon the colonies and form unisexual clusters or gather on nests other than their own, in this case altering the percent presence of the two sexes on the recipient colonies (Turillazzi 1985a). The average female/male ratio at emergence observed during my observation period on this species was 1.27.

Charlotte Samuel made a very detailed study on the presence of males and females on a large number of colonies of *Liostenogaster flavolineata* in peninsular Malaysia and related her observation to annual climate changes. She observed that an average of almost 74 % of the nests in three study periods had no males. The total male population increased from the end of October to the end of December followed by a decrement in two study periods with a minimum in the middle of

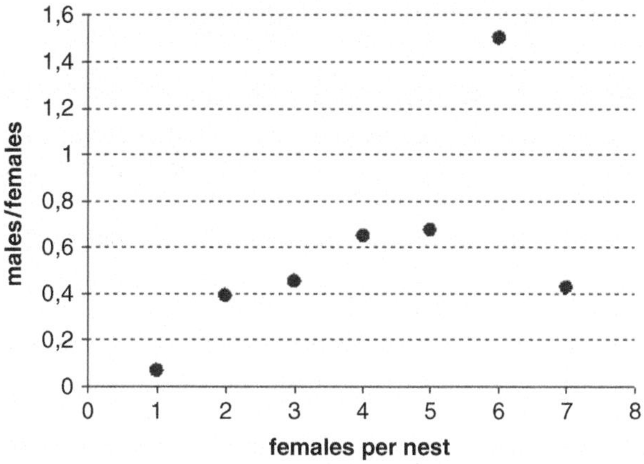

Fig. 4.9 Males and females counted on nests with different number of females in a total of 982 checks on 50 colonies of *L. flavolineata* (data from Samuel 1987)

March and April. The maximum female populations coincided with the drier periods in the various years (Samuel 1987, p. 114 etc.). Between February and July 1982 she performed a total of 982 checks on 50 nests noting the number of females and males present. From her data, reported in the table on page 132 of her unpublished thesis we can construct the graph in Fig. 4.9. It is clear that males are very rare on nests with only two females and their number increases in more mature colonies, similarly to what occurs in *P. nigricans serrei*. In some of these colonies males can even be more numerous than females. The average female/male ratio recorded at emergence on this species was 2.4.

In *Stenogaster concinna* 5 out of 14 individuals, emerged from pupae and prepupae from a sample of 7 nests reared in the laboratory by J.P. Spradbery, were males (Spradbery 1975). In *Eustenogaster fraterna* Gerace (1993) and co-workers found on 20 collected nests 57 adult females and 12 adult males, with a female/male ratio of 4.75. However the individuals emerged in the laboratory from pupae and pre-pupae (of the same nests) were 31 females and 27 males, with a female/male ratio of 1.14: this shows that also in this species nest population is strongly female biased, but not so the overall sex ratio.

In *L. vechti* I found different sex ratios in different nest populations even if the data came from a limited number of colonies. In a locality at 610 m asl. I recorded an average female/male ratio of 4.72 on 22 colonies while this was only 2.17 in 12 colonies collected at about 1,000 m asl (Turillazzi 1990a).

Sex ratio can fluctuate at different times of the year. The most striking case is that observed in species which live in regions with marked seasonality. In 5 nests of *Eustenogaster nigra,* which lives at the northern border of the distribution area of hover wasps, F. Saito and co-workers found an overall number of 24 dormant females and 11 males in winter. As all but two of the females were still un-inseminated, the Japanese entomologists suggest that mating occurs the following spring, after male and female overwintering. In five nests of *E. scitula* discovered at the same time in Central Vietnam, however, the same researchers found 22 dormant females (all but one inseminated) and only one male (Saito et al. 2009).

4.3 Individual Life Histories

In *P. nigricans serrei*, in nine out of nine cases in which I could follow a colony from its foundation, the very first individual to emerge was a female. Seven of these colonies went on producing a second adult that was female in all cases. Only two colonies survived longer, to give birth to a third adult (a female in both nests). Then, the first male emerged in fourth position in one of the two nests while in the other nest a male was only the fifth adult to emerge (Turillazzi 1985a).

From data collected in the field on this and other species we know something about ovarian maturation in females; instead we have no news about the age at which a male becomes sexually mature.

In *P. nigricans serrei*, examining the ovaries of a female less than 45 days old, I found eggs equal to or even larger in size than already laid ones (Turillazzi 1985a); Samuel, for her part, found that solitary foundresses of *L. flavolineata* can sometimes be younger than 16 days (Samuel 1987). In general we can say that ovary size is correlated with age and older females have significantly larger ovaries than younger females.

In *P. nigricans serrei* all females apparently mate roughly between 20 and 50 days after emergence and fertilised females have more developed ovaries than unfertilised ones in the same age range: this means that females with more developed ovaries look for mating possibilities earlier or that mating induces more accelerated ovarian development (Turillazzi 1985a). Samuel dissected 67 *L. flavolineata* females of known age and found that insemination in this species can occur starting from 15 days and up to 63 days after emergence (Samuel 1987, Fig. 1, p. 243; a presumed fertilised 1-day-old female is probably an error). Samuel also found two unfertilised females over 220 days old but with developed ovaries. In any case, in this species fertilised females also present more developed ovaries than unfertilised females of the same age.

In 1989 I proposed a simple graphical model to describe the emergence sequence of individuals in a colony of *P. nigricans serrei* on the basis of data, collected in the field, on emergence and disappearance rates. The model predicted an initial production of three–four daughters before the emergence of the first male and the departure from the colony of the females as soon as their ovaries were fully developed, which was estimated, from field data and dissections, to be on average 80 days (Turillazzi 1989). Actually, on the basis of data recorded from natural colonies for emergence rates (in Y colonies 1 female every 12 days, in M colonies 1 female every 18 days and one male every 17 days and in Ma colonies 1 female every 9 days and 1 male every 12 days) and disappearance rates (in Y colonies 1 female every 25 days, in M colonies 1 female every 25 days and one male every 66 days and in Ma colonies 1 female every 10 days and 1 male every 10 days) for females and males we can obtain a graph of the type in Fig. 4.10. The model describes the development of a hypothetical colony over a period of about 450 days,

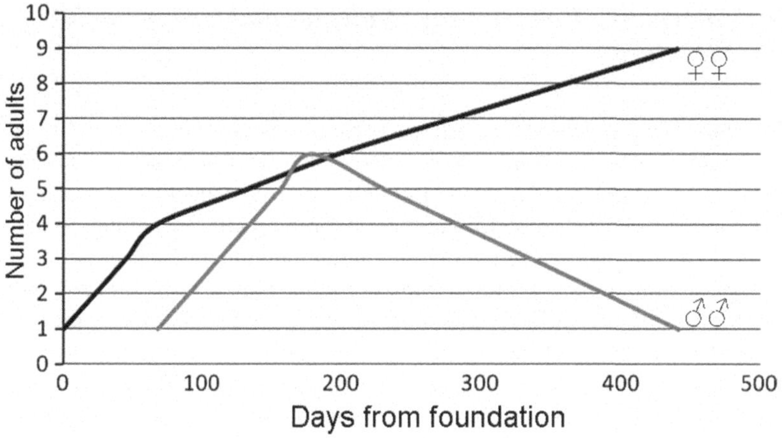

Fig. 4.10 A rough graphical description of the development of an imaginary colony of *Parischnogaster nigricans serrei* (from Turillazzi 1989 re-calculated) (see text)

remaining in the population limits observed for natural colonies, without considering any social factors influencing leaving and staying decisions of females and possible external factors at all, such as episodes of predation, which can affect the life of the colony itself.

Behavioural observations carried out on social interactions and reproductive behaviours of individuals in colonies at various stages demonstrated that different species fundamentally maintain the same characteristics throughout the subfamily. However, various categories of females can be identified according to their particular behaviour in species with societies composed of a higher number of individuals, such as *Liostenogaster*, *Parischnogaster* and, to a lesser extent, *Eustenogaster*. "Foundresses" can be further distinguished in "type I foundress females", individuals which found a new nest or utilise an old nest, and "type II foundress females" that adopt a nest with immature brood abandoned by the previous occupants (Samuel 1987). "Lone females, floaters, or vagrant females" are individuals which are temporary homeless. Nest residents can be further divided into "dominants", the highest ranked individuals on multi-female nests, and "subordinates", individuals which are not at the top of dominance hierarchies.

Later, I shall speak more widely about the characteristics of dominance hierarchies in hover wasps but here suffice it to say that, usually, there is a more or less strict age-based queue for the top rank position. Such a queue system has been described by Hansell et al. (1982), Samuel (1987), Field et al. (1998) and Bridge and Field (2007) for *L. flavolineata*, by Yamane et al. (1983) for *P. mellyi*, by Turillazzi (1985a) for *P. nigricans serrei*, and is defined as "gerontocracy" (Strassmann and Meyer 1983).

4.3.1 Alternative Behavioural Strategies and Group Social Structure

Evolutionary biology postulates that an organism is selected to pass down the highest number of genes similar to its own to the next generation. This can be achieved in two fundamental ways: through direct reproduction, generating the highest number of fertile brood (direct fitness), or indirectly, by helping to increase the number of brood generated by a close kin (an individual which shares a certain percentage of similar genes) (indirect fitness). In the course of its life, any individual can follow different behavioural strategies to maximise its fitness, this will be determined in part by genetic predisposition but, above all, by reactions to environmental situations. Hover wasps are particularly good subjects for studying the life history traits in individuals. This happens because the social constraints are minimal compared with those imposed on a female in a more socially evolved and complex society, such as those that are seen in other social insects. In fact, the colonies of hover wasps do not present rigid castes, and individuals are conditioned in their choices only by the contingent pressures of the social environment around them. Moreover, the relatively stable environments in which these wasps live do not put an end to the colony cycle and make it possible for each emerged female to leave her maternal nest and initiate a new colony (Samuel 1987; Field et al. 1999).

One way of describing the essential characteristics of an animal social system is that proposed by Helms Cahan et al. (2002) which traces the social structure of a species resulting from a series of individual decisions regarding three fundamental points (1) whether or not to disperse from the natal nest/territory, (2) whether to co-breed or not and (3) whether, and to what extent, to provide alloparental care.

The Japanese researcher Soichi Yamane and his colleagues (Yamane et al. 1983) observed that females of *Parischnogaster mellyi* follow three main behavioural options which will condition their future life and their fitness acquisition. They can leave the maternal colony to try to found their own colonies; they can remain on the maternal nest waiting to succeed the dominant female when she fails or try to dethrone her; otherwise, they can resign themselves to act as workers and rear a certain number of individuals which are genetically related to themselves. A similar set of behavioural options was reported for *L. flavolineata* (Hansell et al. 1982; Samuel 1987; Field et al. 1999), *E. calyptodoma* (Hansell 1987), *P. nigricans serrei* (Turillazzi 1985a; 1989). Moreover, Samuel observed that in *L. flavolineata*, females do not limit themselves to just one life strategy but may exercise several of these options over their life time (Samuel 1987).

4.3.2 To Leave or Stay?

The length of time that females remain on a colony probably depends on a number of factors, such as the condition of their reproductive apparatus, the state of the colony and their order of emergence (Turillazzi 1989). For example, according to Samuel (1987), in *L. flavolineata,* females which exhibit a greatly accelerated ovarian development tend to leave their nests early in life, whereas females with retarded ovarian development remain on their natal nest longer, even though any certain cause–effect relation cannot be established. The same author observed that a large proportion of females disperse up to 82 days after emergence. Field et al. (1999) confirmed that in this species young females are more likely to leave than older females, although leaving is possible for females of any age. Most of the leavers spend a period as "nest floaters" before joining other nests (Samuel 1987; Field et al. 1999). In *P. nigricans serrei,* females particularly leave the colonies after the first male emergence (Turillazzi 1989). In *E. calyptodoma* Hansell (1987) found that young females usually leave the maternal nest to found a colony by building a new nest or (in most cases) by usurping an already existing one. Otherwise these females can join other colonies or return to the nest from where they emerged.

In a captive population of *P. mellyi*, Coster-Longman and I found that most of the absences of young females observed in checks on the colonies was not due to the death of the individuals but, actually, to the high rate of leaving and drifting of these individuals (Coster-Longman and Turillazzi 1998). In a group of well-established colonies in an ample tropical room with the temperature kept at approximately 25°C and relative humidity between 60 % and 80 % with a 12:12 L–D period, we followed a total of 65 females (25 already present in the colonies when collected in nature and 40 emerged in captivity) individually for 115 days, with a maximum interval of six days between two consecutive checks. Only six of the first females (24 %) abandoned their nests in the course of observations while 30 of the new borns departed (75 %), the majority (70 %) within 20 days after emergence ($P < 0.001$). Some of the new born individuals which left their nests almost immediately joined a new nest, while the remainder (around 60 %) spent some time (up to 60 days) wandering around. Moreover, some of these individuals changed nests three times or more, with a maximum of six moves recorded. We considered this behaviour a way for new individuals to collect information and to test the situation of the colonial population in the surroundings, searching for opportunities for enhancing their future fitness (Coster-Longman and Turillazzi 1998). The strategies most adopted (out of a total of 47 moves, including multiple ones) was joining another colony (44.8 %), reutilising vacant nests (31.9 %), founding a new nest (8.5 %), usurping other nests (4.26 %), while 4 females finally returned to their original old nests (Coster-Longman and Turillazzi 1998). The movers' choice of nest to join was influenced, in part, by the number of females present, as nests with a higher number of females than the original one were joined in lower measure than expected, considering all the nests available at a given moment. Moreover, unpublished data by Coster-Longman indicated that the choice was apparently not influenced by the status of the immature brood present in the nest (the "value" of which was scored according to the approximate average number of days necessary to complete the respective larval stage, Hansell 1981). The situation described for our captive population is probably also common in nature, and for other species, but its monitoring in the field presents obvious practical difficulties (Field 2008).

Regarding males, in *L. flavolineata* these usually leave the natal nests soon after emergence (Field 2008), while in *P. mellyi* Fanelli et al. (2005) found that males most often remain on the nests from where they emerged.

Thus the decision whether or not to disperse depends on ecological constraints which can prevent, or limit, independent breeding or can enhance, or reduce, direct or indirect benefits of group living (Stacey and Ligon 1991; Helms Cahan et al. 2002).

4.3.2.1 High Costs of Independent Nesting

Availability of Vacant Nests

Jeremy Field and co-workers, in a series of studies, tested the various hypotheses about the possible factors influencing female decisions to remain on or leave their original colonies.

Independent nesting has two main costs: finding a suitable nesting site and the energy for building the nest (Field 2008). In *L. flavolineata* subordinates hardly ever leave their nests even if they have the opportunity to nest independently by adopting a vacant nest at no building cost. Only 5 % of 111 nests in the same nest cluster, made available through elimination of resident females, were adopted by subordinates of other colonies while only 40 % were adopted by female floaters. The nests which were adopted proved to have the most cells and hold more mature brood, which for the most part was not destroyed by the foster females (Field et al. 1998). Solitary nesting in *P. alternata* can be successful only if the foundresses are joined within a few weeks by other females (in this sense it can also be seen as an associative foundation, Turillazzi 1985b) (Bolton et al. 2006).

The high insecurity of new solitary foundations is further supported by survival estimates for dominants, subordinates and single foundresses of *L. flavolineata* obtained with a mathematical model. These suggest that dominants can live longer than subordinates (132 vs. 58 days), while single females have the same life span as subordinates (Shreeves and Field 2002; Cant and Field 2005).

Reproductive Potential

Field and Foster (1999) organised an experiment to check if helper, non-reproductive individuals in multi female colonies of *L. flavolineata* remain in a subordinate position because of their limited egg laying capacity. After 18 days, subordinate females left alone on the nest (after eliminating the dominant individuals) exhibited spermathecae full of sperms and ovarian development similar to those of the previous dominant individuals. According to the authors, these results indicate that the possibility to attain full reproductive potential and initiate a nest alone should not influence any decision by a helper female to stay or to leave (Field and Foster 1999).

Group Size

For a newly emerged female, staying in a large group may have some disadvantages because of the long queue waiting to reach the top of the hierarchy and direct reproduction. Waiting, however can offer its own rewards, i.e. the high number of helpers present when the young female finally does become the dominant (Field et al. 1999), and/or higher labour force to rebuild the nest in case of destruction (Fanelli et al. 2005). However, Field et al. (1999) found no effect of the size of the group on the decision of the females of *L. flavolineata* to stay and help or to leave and visit (and eventually adopt) nests experimentally made vacant.

Position in the Queue

Offspring production increases linearly with group size in *L. flavolineata*; but groups are limited in size and this means that the chances of inheriting a nest with a good group of helpers are particularly low. Direct fitness seems a relatively large component of total fitness in these wasps (Field 2008) and the disadvantages of being the last in a long queue could be only partially compensated by being in a larger and more productive group. However it is hard to test experimentally the maximum queue length above which a female would be better off leaving (Field 2008).

Fitness Insurance

Another good reason for staying in a group is that your brood will be cared for in case you disappear. In *L. flavolineata* Field et al. (2000) tested if females in large groups had the possibility of saving their investment in immature brood, even after their disappearance, if at least part of the larvae and pupae were reared and protected by other females remaining on the nest (assured fitness return). The authors found that most of the investment made by these individuals (experimentally eliminated from their colony) is preserved, as most of the large larvae and pupae are brought to emergence by their nestmates, with the only loss of small larvae, which are probably used as food for the larger ones. Helpers in *L. flavolineata* are thus favoured by assured fitness return (Strassmann and Queller 1989; Queller 1989; Gadagkar 1990) with respect to lone foundresses unless the latter can invest 1.7 times more, but observational data indicate that the investment (in terms of immature brood) for the two categories of females is almost the same (Field et al. 2000).

Again, in *L. flavolineata*, also due to the very long period of immature development, only 50 % of independent females can hope to live long enough to see the emergence of their own brood. Although a helper has the same life expectancy, her investment can in part be saved by the group, which will in turn have more chance to survive thanks to her help (survivorship insurance) (Reeve 1991; Queller 1996; Nonacs 1991).

4.4 Colonial Genetic Structure

4.4.1 Adult Female Relatedness

Owing to the frequent disappearance of females from *L. flavolineata* and *E. calyptodoma* colonies, Samuel (1987) and Hansell (1987) already concluded that females which emerge on the nests are not usually the daughters of the dominant females they find on emergence. When biomolecular techniques for genetic analyses became available, relatedness between female nestmates was determined for four species (Table 4.2). The very first method used was based on the comparison of allozymes and was performed on colonies of *L. flavolineata* and *P. alternata* (Strassmann et al. 1994). The analyses for *L. flavolineata* were later repeated with techniques based on DNA microsatellite markers (Sumner and Field 2001) employed by Sumner (1999) and Sumner et al. (2002) who obtained much higher values of intra-colony female relatedness in three different populations of these wasps.

Similar values have been obtained for *P. alternata* with allozymes (Strassmann et al. 1994) and with DNA MS markers (unfortunately unpublished to date) by Bolton et al. (2006). In *P. alternata* Bolton et al. (2006) found that females were not all sisters, even though in 12 out of 17 colonies examined the relatedness between adult females was not significantly different from 0.75.

In the common species *P. mellyi* Fanelli et al. (2008) found an average relatedness of 0.45, but an even lower relatedness (0.33) had been observed in an undisturbed set of colonies in a previous research (Fanelli et al. 2005).

Landi et al. (2003) reported an average relatedness of 0.43 for female nestmates in *E. fraterna*, a mountain species studied at Bukit Fraser in Peninsular Malaysia.

These values are typical for colonies of primitively eusocial insects and individuals in hover wasp colonies turn out to be a mixture of sisters, aunt–niece, mother–daughter or cousins (Field et al. 2006).

The only values which are remarkably different are those reported by Strassmann et al. (1994) for *L. flavolineata* (0.22), obtained with allozyme comparisons and that for *P. mellyi* (0.33) reported by Fanelli et al. (2005). In a recent review, Field (2008) practically discards the first, opposing this very low value with the higher value given by Sumner (1999) (0.45) measured on a sample of colonies from the same population as that of Strassmann et al. (1994). Actually

Table 4.2 Values for average intra-colony relatedness of females belonging to 4 species

Species	*N* colonies	Average relatedness	SD	Method	LOC	References
L. flavolineata	38	0.22	0.10	Allozymes	Gombak	Strassmann et al. (1994)
L. flavolineata	27	0.52	0.05	DNA MS	Gombak	Sumner et al. (2002)
L. flavolineata	?	0.45	0.10	DNA MS	GTE	Sumner (1999)
L. flavolineata	?	0.46	0.08	DNA MS	BF	Bridge (2005)
P. alternata	22	0.56	0.19	Allozymes	Gombak	Strassman et al. (1994)
P. alternata	17	0.46	0.054	DNA MS	BF	Bolton et al. (2006)
P. mellyi	19	0.33	0.05	DNA MS	Gombak	Fanelli et al. (2005)
P. mellyi	15	0.45	0.05	DNA MS	Gombak	Fanelli et al. (2008)
E. fraterna	17	0.43	0.13	DNA MS	BF	Landi et al. (2003)

Gombak Gombak Valley; *GTE* Genting Sempah; *BF* Bukit Fraser (all localities from Peninsular Malaysia)

there is no apparent reason not to accept both estimates as good, given also the relevant number of colonies of the sample of Strassmann et al. (1994). Concerning *P. mellyi*, Fanelli et al. (2005) explain the low estimate (0.33), obtained on colonies collected from the same place during subsequent sampling, which gave a significantly higher value (0.45), by the fact that the first sample was actually composed of un-manipulated colonies, while the second one was performed on colonies where the females had previously been marked. Fanelli and co-workers suggested that the marking procedure induced the more vagrant, and probably, less related females to abandon the nests after they had been disturbed.

Estimates by Strassmann et al. (1994) on males present on *L. flavolineata* nests indicate that these probably emerged on the same colonies on which they had been found. These authors conclude that the low relatedness levels observed in this species cannot be explained by indirect fitness benefits for the scarcely related, non-reproductive helpers but, rather, as the result of responses to ecological factors.

4.4.2 Brood Relatedness

Strassmann et al. (1994) found that in the few colonies of *L. flavolineata* where they had a sample of female pupae, these were in all probability full sisters. On the contrary, this was not the case when considering the female brood of all ages (Sumner et al. 2002) both in this species and in *P. alternata* (Bolton et al. 2006).

4.5 To Co-breed or Not to Co-breed?

In earlier studies on the composition of colonies of various hover wasps a relatively high number of females with developed ovaries were found at dissection. In populations of some species of *Parischnogaster*, for example, percentages of potential egg layer females (females with eggs in their ovaries similar or larger in size than laid ones) are around 38 % in *P. nigricans serrei* and 36 % in *P. mellyi* and rise to over 50 % in *P. alternata* and up to almost 70 % in *P. striatula*. Percentages of fertilised females were around 85 % in these latter species and around 69 % and 63 % in the first two species (Turillazzi 1990c). Considering the average percentages per colony I obtained lower but still high values: 53 % in *P. striatula*, 51 % in *P. alternata* and 37 % in *P. nigricans serrei* (Turillazzi 1989).

In *L. vechti* I found average percentages of 28 % potential egg layers per colony; interestingly, however, in colonies of a cluster of nests found at a site 600 m asl the number of PEL females per colony was almost constant around 1–2, regardless of the total number of females present, while it rose with the total number of females in the colonies of a cluster found at an altitude of 1,000 m asl. Differences in the correlation indexes were highly significant showing that even the number of egg layers per colony can be affected by environmental factors (see Fig. 4.11) (Turillazzi 1990a).

However, notwithstanding the presence in colonies of various species of several potentially reproductive individuals, genetic analyses of immature brood pointed to the fact that in all the species examined most of the eggs are produced by only one single mated female.

In *L. flavolineata*, for example, Sumner et al. (2002) showed that the dominant female was more closely related to the eggs (male and female) than her subordinates in 11 out of 13 examined nests (skew range 0.68–1). Moreover 90 % of the male eggs were laid by the dominant female and overall male production was shared between the dominant and the highest ranked subordinate female that, in any case, did not lay female eggs. Sumner et al. (2002) concluded that the high skew in *L. flavolineata* may result from ecological constraints and the relatively high probability of a subordinate inheriting the top of the hierarchy and egg-laying position.

In *E. fraterna* every nest out of 12 examined had only one single mated female which usually produced most of the brood (Landi et al. 2003).

In *P. mellyi* in only four out of 19 colonies examined, direct reproduction was shared between two females, but the dominant one laid more than 92 % of the total eggs and 98 % of the female eggs, while the subordinate egg layer laid either an equal or a larger proportion of male eggs than the dominant. This occurred only when egg layers were unrelated (Fanelli et al. 2005). Shared reproduction between unrelated females, probably joiners or usurpers, can in part explain the low relatedness observed in this species (0.33) (Fanelli et al. 2005).

In *P. alternata* the production of both female and male offspring was almost completely monopolized by a single dominant female (Bolton et al. 2006).

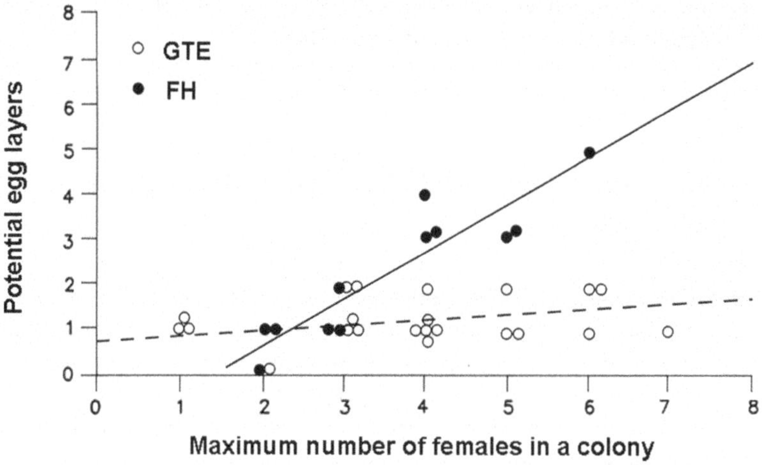

Fig. 4.11 Number of potential egg layer females and total number of females per colony registered in two clusters of nests of *Liostenogaster vechti* at different altitudes (GTE: 650 m asl; FH: 1,000 m asl, Central Mountain Range, Peninsular Malaysia). From Turillazzi 1990a re-drawn

4.5.1 Factors Influencing Reproductive Partitioning

How can stable strategies evolve to produce variations in reproductive partitioning in animal societies, the so-called "reproductive skew"? Since in all the species of hover wasps studied so far, females are behaviourally flexible they are particularly good for testing models of reproductive skew (Sumner et al. 2002; Fanelli et al. 2005).

Possible variables which can explain reproductive partitioning are relatedness, body size, group dimensions and productivity; however, not one of these parameters is correlated with the skew observed in *L. flavolineata* by Sumner et al. (2002).

Reproductive skew across colonies was not significantly correlated with relatedness, body size and productivity in *P. mellyi* but Fanelli and co-workers found a higher share of reproduction in colonies where the nests had a higher number of empty cells (Fanelli et al. 2005, 2008).

In *P. mellyi* Fanelli et al. (2008) used aggression as a key parameter to test how some variables can affect the reproductive partitioning of females in a colony. They showed that lower reproductive skew is associated with higher levels of aggression by the subordinates and, noting that no currently available model is supported by their results, suggested that a mechanism, based on aggression, regulating the sharing of reproduction could be applied to all primitively eusocial wasps. The same authors also suggest, with a life history-based hypothesis, that the presence of unrelated females in a colony without any hope of immediate direct reproduction is better explained with the hypothesis that subordinates, considering the strong constraints against independent breeding, still remain because of future chances to become dominant on that colony (Fanelli et al. 2008).

4.6 Helping and Alloparental Care: To Collaborate or Not to Collaborate?

In *E. calyptodoma* Hansell (1987) reports that, even "in the absence of any obvious dominance relationship", the younger individuals in colonies formed by two females are those that perform most of the foraging while the dominant, older females mainly spend their time in the nest guarding the nest entrance.

In *E. fraterna* division of labour is somewhat different from that observed in other species, even in the same genus, because young females with undeveloped ovaries usually work less hard and take fewer risks than the dominant females that continue to forage and build the nest, even after the emergence of the first adult brood. Francescato et al. (2002a) observe that, in this case, young females give an important contribution anyway, as they can defend the nest from usurpation from alien conspecifics. This situation could be related with the high mountain environment where *E. fraterna* lives. In this respect, it would be interesting to compare the division of labour among females in colonies in nest sites at different altitude in those species which present a wider altitudinal distribution, such as *P. striatula*.

Field et al. (2006) observed that by working hard a subordinate raises her indirect fitness component but decreases the possibility of gaining a high direct reproduction. In colonies of *L. flavolineata*, they manipulated two fitness components in subordinate females in the hierarchical queues composed by four individuals. In the first experiment, they eliminated individuals in front of rank-three females and, afterwards, compared the helping efforts of these ones with those of individuals, the queues of which had been manipulated by eliminating the lower (4th) ranked nestmates. In the other experiment, the groups of second rank (beta) females were reduced by eliminating lower rank individuals (3rd and 4th), then the helping efforts of the beta females were compared with those of beta females the group of which had remained intact (see Field et al. 2006, Fig. 2). In the first experiment, the experimental rank-three females worked less than the controls. In the second experiment, the authors found that the focal (rank-two) females worked harder on colonies where the group size had been reduced. These results were in line with the prediction of the authors that the social environment (in this case position in the hierarchy and group size) provides a good indication about the future fitness perspectives for a member of the group, giving it the possibility to mould, in some way, its energy investment, even without any indications about relatedness (Field et al. 2006).

4.6.1 Dominance Hierarchies

As we have seen, sharing reproduction and division of labour, including efforts to adjust the related energetic costs in order to maximise individual fitness, are regulated by dominance hierarchies. These can be seen as sets of social regulatory mechanisms which can be influenced by particular characteristics of the individuals involved in the game (body size and fighting ability, physiological predisposition, particular phenotypic traits mainly correlated with health conditions, etc.).

Yoshikawa et al. (1969) were the first to report an example of dominance hierarchy in hover wasps. They observed and described it in a six female colony of one species of *Parischnogaster* (probably *jacobsoni)* studied at the Gap (Central Peninsular Malaysia). The authors found a definite linear dominance order in the first three ranks, but it was not possible to rank the last three females (two of which unfertilised) which were often absent from the nest. The top rank female, the one with the most developed ovaries, was almost always present on the nest, limiting her activities to resting and patrolling and dominating (in particular the rank-three female).

4.6.1.1 Types of Dominance Hierarchy

Gerontocracy is a conventional assessment of dominance hierarchies where older individuals in a group are dominant over younger ones (Bridge and Field 2007). Concerning insects, the phenomenon was first described by Pardi in the colonies of the paper wasp *P. dominula* where older workers get to the top of the hierarchy after the disappearance of the foundress female (Pardi 1946). The system probably evolved because older individuals can also be "better", maybe due to their greater experience in managing the colony (Bridge and Field 2007).

Gerontocracy seems to be widespread in hover wasps and the seeds of an organisation of dominance hierarchies based on relative age of the individuals can be observed even in those species which have quite simple social structures, such as those of *Stenogaster*, *Anischnogaster* and *Metischnogaster,* where the mother is dominant over the young daughters before they leave the nest.

In more socially complex hover wasps, ranks in the social hierarchy of a colony are mainly determined by the relative age of the individuals (Hansell et al. 1982). In *L. flavolineata* the oldest female in a colony is in 87 % of cases the dominant one (Samuel 1987); the small size of the colonies and lack of an inactive season in the tropical environment mean that individual reproduction, once the top position in the hierarchy is reached, is the largest component of life fitness in females of this species (Bridge and Field 2007).

In his study on social behaviour of *P. mellyi*, Hansell (1983) stressed the fact that, in a given colony, generally neither a simple linear hierarchy nor a unique dominant female was apparent; but Yamane et al. (1983), studying the same species in Sumatra, found that there was a clear dominance relation between female nestmates, regardless of whether these originally came from the same or different nests. The last authors could not ascertain, however, if daughters or step-daughters could replace the top dominant individuals at a certain time or had to spend all their lives as workers.

When, together with Leo Pardi, we studied *Parischnogaster nigricans serrei* in West Java, we realised that individuals were arranged in a linear hierarchy of the type observed in *Polistes* wasps. Dominant females behaved aggressively towards their nestmates whilst the latter showed clear signs of submission and avoidance with respect to higher ranked individuals. Dominance and subordinance acts were easier to distinguish than in other species simply because of the particular shape of the nest, as individuals have to confront each other every time they walk on the long, linear series of cells (Turillazzi and Pardi 1982). In one case we could also observe substitution of the top ranked female in a colony by the second ranked one. In this species, as well as in the colony observed by Yoshikawa et al. (1969) the relative positions in the hierarchy could be established by the simple observation of aggressive interactions even if these were actually milder with respect to those reported for other social wasps (Turillazzi and Pardi 1982).

In *P. alternata* intra-colonial aggressive interactions seem more diluted, but hierarchies can be recognised, especially for the first positions (Turillazzi 1986). In 10 out of 12 observed nests of this species the dominant was the female that had been resident for longer, but only in 3 out of 7 colonies the dominant female was also the largest individual (Bolton et al. 2006).

Queue Jumpers

Hierarchies appear quite stable in *L. flavolineata* as only 13 % of the observed individuals jumped the age queue (Samuel 1987), and these individuals are usually those that have worked less than others (Bridge and Field 2007). The same authors also observed that some queue jumpers actually work less hard than expected for their position in the queue and indicated the possibility that these individuals perform a frequency-dependent cheating strategy.

It is clear, at this point, how important it can be for helper individuals to know their position in the dominance queue in order to adjust their behavioural strategies but, at present, we still do not know the mechanisms, or cues, which females use to do so (Bridge and Field 2007).

4.6.1.2 Division of Labour

In *S. concinna*, Spradbery (1975) affirms not to have found any evidence of co-operative behaviour among adults present on the nests and he states that this species displays the most primitive sub-social condition. In one two-female colony, he recorded that one individual was fertilised, with developed ovaries, and the other was instead unfertilised with developing oocytes. The same situation can be found in other genera. In *Anischnogaster* sp.A only one two-female colony was observed and the younger female was the foraging individual (86 % of the time absent from the nest), while the older female remained almost constantly on the nest (with only 0.3 % of the time absent) (Turillazzi and Hansell 1991). Similarly, in *Metischnogaster drewseni* a young female begins to forage after some days from emergence but remains as a helper on the nest for only a brief period before leaving definitively (Turillazzi 1990b).

In *L. flavolineata* dominant females were found on the nest in 98 % of the censuses performed on colonies while the highest rank subordinates were observed only in 69 % of cases (Hansell et al. 1982). Helping activities are carried out from the very first week after emergence by young, non-dominant individuals that are engaged in foraging, production of abdominal secretion and feeding the larvae. By 2 weeks, they also begin to collect pulp and assist in nest building. Only older females seem to take part in nest defence (Samuel 1987).

Hansell (1983) noted a distinct difference in the presence on the nest by females of *P. mellyi* in Thailand, with a senior female spending almost all her time on the nest and resting in a position above the other nest mates. He observed also that some females can specialise in a particular helping activity, for example, foraging and collecting building material. Yamane et al. (1983), studying colonies of the same species in West Sumatra, observed common inter-colony drifting and the formation of stepmother–stepdaughter associations in most nests. Division of labour was quite clear in colonies of these wasps; apart from the distinctive differences in internal and external activities, dominant and subordinate females participated to the same extent to immature brood rearing, but dominants were the only ones to initiate and enlarge cells, to lay eggs and to perform most of the aggressive acts (Yamane et al. 1983, 1990).

In *P. nigricans serrei* both the degree of ovarian development and division of labour in multi-female colonies were correlated to hierarchical rank, with the leader female almost always resident on the nest, where she alternated long periods of immobility with cell inspection, patrolling, egg laying and aggressive flights against intruders approaching the colony. Subordinate females were more often absent and performed the bulk of the foraging, contributing also to most nest activities, including larval nutrition and ant guard construction (Turillazzi and Pardi 1982).

In *P. jacobsoni* I compared the helping behaviour of eleven females with ready-to-be-laid eggs in their ovaries (PEL) and other females ($n = 17$) and I noticed that PEL tend to leave the nest less and perform more dominance acts. However, I did not find any significant differences in other types of behaviour (Turillazzi 1988).

In *P. alternata* a colony of three females was observed for 12 h consecutively and a clear division of labour was recorded, notwithstanding all three individuals had developed ovaries. The dominant female (that with the most developed ovary) remained almost continuously on the nest, patrolling and defending it from intruders, while the female with the less-developed ovaries was the one that remained most time out of the nest. The third female showed intermediate activity values between the other two (Turillazzi 1986). In this species it is also common to find pairs of females on associative foundations; their behavioural and anatomical characteristics, however, are very similar, including ovarian development. As we have seen, it seems that the advantages of this kind of association in this species, which forms huge clusters of colonies in suitable nesting sites, are mainly the greater efficiency with which two females can defend the nest from conspecific females or from females of *L. flavolineata*, coming from neighbouring colonies or wandering among the cluster. Such females are usually active in stealing material and even Dufour's gland secretion and eggs from other nests (Turillazzi 1986; Turillazzi et al. 1997).

Cronin and Field (2007) examined nest defence behaviour in females of *L. flavolineata* colonies against con-specific intruders and observed that high rank individuals defend more than low rank individuals, while the latter contribute more to foraging. The same pattern of division of labour is reported in some species of *Polistes*, such as *P. fuscatus* by Gamboa et al. 1992. In *P. dominula* dominant females are the ones which defend their colonies from the attacks of the social parasite *P. sulcifer* most aggressively (Turillazzi et al. 1990).

In theory, subordinates should work less if they stay in a large group and have more probability of reaching the top of the queue in a short time, but they work more if the size of the group is small, since they then stand to inherit less and so have less to lose by helping (Cant and Field 2001, 2005). In experiments manipulating queue length and group size (see Sect. 4.6), Field et al. (2006) ascertained that subordinates can adjust their helping (foraging) efforts depending not only on their position in the dominance queue but also on the size of the group itself (in line with the prediction of the model elaborated by Cant and Field 2001, 2005). In fact, individuals near the top of the queue work less but increase their efforts again if their group is artificially reduced.

In *E. fraterna* Francescato et al. (2002) hypothesise that subordinates work less to increase the egg-layer mortality in order to inherit the nest and the immature brood present earlier, as suggested by Queller (1996).

4.6.1.3 Other Factors Influencing Rank Determination

Size

In *P. nigricans serrei* fertilised females have significantly larger heads and better-developed ovaries than unfertilised females (Turillazzi and Pardi 1982). Relative size, however, as well as relatedness, does not seem to influence the rank order of nest mates in *L. flavolineata* (Hansell et al. 1982; Samuel 1987; Field et al. 1999; Sumner et al. 2002; Bridge and Field 2007).

Physiology

A physiological predisposition to function as egg layers in certain females was excluded by Field and Foster (1999) who forced low-ranked helper females to remain alone on their nests in order to ascertain that after almost three weeks of isolation most of them had mated and succeeded in obtaining fully developed ovaries. The authors stated that these results do not support the subfertility hypothesis of West-Eberhard (1978) as they demonstrate that helper females are not physiologically constrained from immediate direct reproduction and that only the social environment prevents them from fully exploiting their reproductive potential. However, the authors themselves acknowledged that their experimental females were somewhat advantaged with respect to lone foundresses as they had an already initiated nest and various larvae to utilise as possible food sources. In this way they could have saved the energy required for reaching their full reproductive potential.

Manipulation

Strong influence of the behaviour of other females (and possibly by males) could also add to the factors regulating and maintaining the dominance queue; even if this is quite probable, experimental evidences is still lacking for any species. In various social wasps, for example, oophagy by dominant individuals of the eggs of subordinates is usually reported. In the case of hover wasps, oophagy has hardly ever been observed but could be higher than expected (Turillazzi and Pardi 1982).

Endocrine Status and Hierarchy

The juvenile hormone is important in defining and regulating the ovarian development of adult insects. This hormone is produced by the Corpora Allata, two endocrine glands situated just behind the brain. In some insects, including some social wasps, it has been discovered that the size of the C.A. in a female is highly related with the development of her ovary and the hierarchical rank she occupies in her colony (cf. Röseler et al. (1980) and Turillazzi et al. 1982, for *Polistes dominula*) In an unpublished research on *P. alternata*, I measured the size of the ovaries and the volume of the C.A. from females belonging to 12 colonies with a number of residents ranging from 2 to 6. There was no statistical correlation between the size of CA and that of the ovary taking all the females together, but there was a significant positive relation between the position of a female in a given colony for ovarian development and the position for the size of her C.A. (Spearman ro $= 0.520$, $P = 0.001$, $N = 35$) and for that of C.A. and that of body size (Spearman ro $= 0.340$, $P = 0.045$, $N = 35$). In 6 colonies out of 12, the female with the largest C.A. was also the one with the most developed ovaries. These results wait confirmation but can be an indication of a significant influence of social environment on the reproductive physiology of these wasps. In fact, according to Röseler et al. (1980) the effect of a dominant is to inactivate the subordinates' C.A. which in turn inhibits oogenesis.

Status Signals

Recently Tibbetts and co-workers (Tibbetts 2002; Tibbetts and Dale 2004; Tibbetts et al. 2010) discovered that clypeal marks in female *Polistes* wasps may have some importance in determining the social rank of an individual, acting as a sort of status signal and a mark of genetic goodness to be shown in various types of contexts. These clypeal marks would imply social costs for their bearers which would be particularly high in case of mismatches between signal and behaviour (Tibbetts et al. 2010). The validity and the actual characteristics of these signs are still debated as different populations of *P. dominula* from that analysed by Tibbetts present very low percentages of wasps with clypeal marks (Cervo et al. 2008 on an Italian population). The results obtained by Zanette and Field (2009) on another population of *P. dominula* (in Southern Spain) are consistent with the possibility that clypeal marks represent a signal of quality, but they cannot be essential cues for the establishment of dominance hierarchies. Females of various species of *Liostenogaster* have evident facial markings constituted by dark brown colouration of the posterior part of the clypeus, and on the front (see Fig. 5.3). Males of the same species have more yellow colouration of the same parts. Analysis of a population of colonies of one species of *Liostenogaster* recently performed by my group gave interesting results. In *L. vechti* we discovered a strong positive correlation between the size of facial brown markings of the females belonging to a given colony and their ovarian development, and in practice with their rank position. Behavioural experiments demonstrated that females with larger facial markings are more fiercely attacked by resident females from alien colonies when presented near their nests. Our hypothesis is that these markings represent a sort of status signal which is used during approaching flights (or landings) of wandering females to test the reactions of resident individuals and the prospects they have of joining, or not the colony with a good position for direct reproduction (Baracchi et al. 2012; see also Chap. 5).

4.7 The "Decision Tree" and "Social Trajectories" in Hover Wasps

According to Helms Cahan et al. (2002) a "decision tree" is the representation of the range of social decisions which are available to each individual of a given species during its life time; a "social trajectory", on the other hand, represents the set of decisions for any given individual. From the previous review, it seems obvious that even if the main decisions can be categorised also for hover wasps, the characteristics of their biology make it quite complex to trace the social trajectories available for one species, also given the fact that the decisions may not always be absolute or dichotomous. Moreover the ecological conditions for any individual can change during his/her life and previous decisions taken can, by that time, make a fixed trajectory optimal. As I have stressed on other occasions, this mainly depends on the wide range of ecological constraints that differ for different species. For example, the decision to disperse or to remain on the nest can be more open in species of the genera *Parischnogaster* or *Liostenogaster* but more limited in species of other genera, such as *Stenogaster*, *Anischnogaster* or *Metischnogaster*.

The high percentages of females with developed ovaries that can be found in certain species where, in any case, only one female lays eggs at a time, is another sign of the uncertainty of clear adaptive strategies offered to the females. Probably more detailed observations on the biology of these species and more experimental work will contribute to understanding why some females develop ovaries but do not reproduce.

Field (2008) affirms that gerontocracy is, by itself, evidence contrary to the subfertility hypothesis (West-Eberhard 1978) as "today's "subfertile" helpers are tomorrow's dominant egglayers"; however, the widespread tendency to escape the rules found in many species, including *L. flavolineata*, shows that other contingent factors may influence the ability of a given female to exploit her reproductive potential. On this line Fanelli et al. (2008) observe that an individual's reproductive potential and disposition to dominate may be determined by a great number of variables such as age, nutritional status, health and contingent decisions about which behavioural strategies to adopt. Moreover, contingent environmental accidents and situations can heavily influence the life trajectories of single individuals. This means that accurate predictor variables for dominance cannot be available, especially for hover wasps where behavioural strategies are evidently so varied with respect to those of more evolved social insects (Fanelli et al. 2008, personal observation).

I shall return to these arguments when I try to face the problem of the evolution of social behaviour in these wasps.

References

Baracchi D, Hashim R, Teseo S, Turillazzi S (2009) Basic social biology and nest architecture of *Liostenogaster topographica* Turillazzi 1999 (Hymenoptera Stenogastrinae). Trop Zool 22:15–25

Baracchi D, Petrocelli I, Cusseau G, Pizzocaro L, Teseo S, Turillazzi S (2012) Facial markings in the hover wasps: quality signals and familiar recognition cues in two species of Stenogastrinae. Anim Behav in press

Barthélémy C (2008) A provisional guide to the social vespids of Hong Kong (Hymenoptera Vespidae). On line: http://insectahk.com/book_page.html

Barthélémy C (2009) Notes on biology and nests of a hover wasp, Eustenogaster nigra (Vespidae: Stenogastrinae), in Hong Kong. Hong Kong Entomol Bull 1:27–31

Bolton A, Sumner S, Shreeves G, Casiraghi M, Field J (2006) Colony genetic structure in a facultatively eusocial hover wasp. Behav Ecol 17:873–880

Bongiovanni V (1998) L'architettura del nido ed il materiale da costruzione nelle vespe Stenogastrinae (Hymenoptera: Vespidae). Msc thesis, Università di Firenze, Italy

Bridge CAL (2005) Rank and inheritance in a facultatively eusocial hover wasp. PhD Thesis, University College London, London

Bridge C, Field J (2007) Queuing for dominance: gerontocracy and queue-jumping in the hover wasp *Liostenogaster flavolineata*. Behav Ecol Sociobiol 61:1253–1259

Cant MA, Field JP (2001) Helping effort and future fitness in cooperative animal societies. Proc R Soc B 268:1959–1964

Cant MA, Field JP (2005) Helping effort in a dominance hierarchy. Behav Ecol 16:708–715

Cervo R, Dapporto L, Beani L, Strassmann JE, Turillazzi S (2008) On status badges and quality signals in the paper wasp *Polistes dominulus*: body size, facial colour patterns and hierarchical rank. Proc R Soc B 275:1189–1196

Coster-Longman (1998) Ecological factors in the agglomeration and evolution of sociality in the Stenogastrinae (Hymenoptera Vespidae). PhD thesis, Università di Firenze, Italy

Coster-Longman C, Turillazzi S (1998) Leaving the nest and related behavioural strategies in a captive population of *Parischnogaster melly*i (Stenogastrinae, Vespidae, Hymenoptera). Insect Soc Life 2:37–44

Coster-Longman C, Landi M, Turillazzi S (2002) The role of passive defense (selfish herd and dilution effect) in the gregarious nesting of Liostenogaster wasps (Vespidae Hymenoptera Stenogastrinae). J Insect Behav 15:331–350

Cronin AL, Field J (2007) Rank and colony defense against conspecifics in a facultatively eusocial hover wasp. Behav Ecol 18:331–336

Cronin AL, Bridge C, Field J (2010) Climatic correlates of temporal demographic variation in the tropical hover wasp *Liostenogaster flavolineata*. Insect Soc 58:23–29

Fanelli D, Boomsma JJ, Turillazzi S (2005) Multiple reproductive strategies in a tropical hover wasp. Behav Ecol Sociobiol 58:190–199

Fanelli D, Boomsma JJ, Turillazzi S (2008) Subordinate wasps are more aggressive in colonies with low reproductive skew. Anim Behav 75:879–886

Field J (2008) The ecology and evolution of helping in hover wasps (Hymenoptera: Stenogastrinae). In: Korb J, Heinze J (eds) Ecology of social evolution. Springer, Heidelberg, Berlin

Field J, Foster W (1999) Helping behaviour in facultatively eusocial hover wasps: an experimental test of the subfertility hypothesis. Anim Behav 57:633–636

Field J, Foster W, Shreeves G, Sumner S (1998) Ecological constraints on independent nesting in facultatively eusocial hover wasps. Proc R Soc B 265:973–977

Field J, Shreeves G, Sumner S (1999) Group size, queuing and helping decisions in facultatively eusocial hover wasps. Behav Ecol Sociobiol 45:378–385

Field JP, Shreeves G, Sumner S, Casiraghi M (2000) Insurance-based advantage to helpers in a tropical hover wasp. Nature 404:869–871

Field J, Cronin A, Bridge C (2006) Future fitness and helping in social queues. Nature 441:214–217

Francescato E, Massolo A, Landi M, Gerace L, Hashim RB, Turillazzi S (2002) Colony membership, division of labour and genetic relatedness among females of colonies of *Eustenogaster fraterna* (Hymenoptera, Vespidae, Stenogastrinae). J Insect Behavior 15:153–170

Gadagkar R (1990) Evolution of eusociality: the advantage of assured fitness returns. Philos Trans R Soc Lond B 329:17–25

Gamboa GJ, Wacker TL, Duffy KG, Dobson SW, Fishwild TG (1992) Defence against intraspecific usurpation by paper wasp cofoundresses (*Polistes fuscatus*, Hymenoptera: Vespidae). Can J Zool 70:2369–2372

Gerace L (1993) Caratteristiche della popolazione coloniale e descrizione del nido di Eustenogaster fraterna (Bingham) (Hymenoptera Vespidae). MSc thesis, Università di Firenze, Italy

Hansell MH (1981) Nest construction in the subsocial wasp *Parischnogaster mellyi* (Saussure) (Stenogastrinae Hymenoptera). Insect Soc 28:208–216

Hansell MH (1982) Brood development in the subsocial wasp *Parischnogaster mellyi* (Saussure) (Stenogastrinae Hymenoptera). Insect Soc 29:3–14

Hansell MH (1983) Social behaviour and colony size in the wasp *Parischnogaster mellyi* (Saussure), Stenogastrinae (Hymenoptera, Vespidae). Proc Konin Neder Akad Wetens Ser C 86:167–178

Hansell MH (1986) Colony biology of the stenogastrine wasp *Holischnogaster gracilipes* (Van der Vecht) on Mount Kinabalu. Entomol Mon Mag 122:31–36

Hansell MH (1987) Elements of eusociality in colonies of *Eustenogaster calyptodoma* (Sakagami & Yoshikawa) (Stenogastrinae, Vespidae). Anim Behav 35:131–141

Hansell MH, Samuel C, Furtado JI (1982) *Liostenogaster flavolineata*: social life in the small colonies of an Asian tropical wasp. In: Breed MD, Michener CD, Evans HE (eds) the Biology of Social Insects. Westview Press, Boulder

Helms Cahan S, Blumstein DT, Sundström L, Liebig J, Griffin A (2002) Social trajectories and the evolution of social behavior. Oikos 96:206–216

Iwata K (1967) Report of the fundamental research on the biological control of insect pests in Thailand. II. The report on the bionomics of Aculeate wasps bionomics of subsocial wasps of Stenogastrinae (Hymenoptera, Vespidae). Nat Life SEA 5:259–293

Jeanne RL (1979) A latitudinal gradient in rates of ant predation. Ecology 60:1211–1224

Krombein KV (1991) Biosystematic studies of Ceylonese wasps. XIX: Natural History Notes in Several Families (Hymenoptera, Eumenidae, Vespidae, Pompilidae, Crabronidae). Smithsonian Contrib Zool 515:1–41

Landi M, Queller DC, Turillazzi S, Strassmann JE (2003) Low relatedness and frequent queen turnover in the stenogastrine wasp *Eustenogaster fraterna* favor the life insurance over the haplodiploid hypothesis for the origin of eusociality. Insect Soc 50:262–267

Matsuura M, Yamane SK (1990) Biology of the vespine wasps. Springer, Berlin

Nonacs P (1991) Alloparental care and eusocial evolution: the limits of Queller's head-start advantage. Oikos 61:122–125

Ohgushi R, Sakagami SF, Yamane S (1990) Nest architecture of the stenogastrine wasps: diversity and evolution (Hymenoptera, Vespidae). A comparative review. In: Sakagami SF, Ohgushi R, Roubik DW (eds) Natural history of social wasps and bees in equatorial Sumatra. Hokkaido University Press, Sapporo

Pardi L (1946) Ricerche sui Polistini VII. La "dominazione" ed il ciclo ovarico annuale di *Polistes gallicus* (L.). Boll Ist Entom Univ Bologna 15:25–84

Queller DC (1989) The evolution of eusociality: reproductive head starts of workers. Proc Natl Acad Sci USA 86:3224–3226

Queller DC (1996) The origin and maintenance of eusociality: the advantage of extended parental care. In: Turillazzi S, West-Eberhard MJ (eds) Natural history and the evolution of paper wasps. Oxford University Press, Oxford

Reeve HK (1991) Polistes. In: Ross KG, Matthews RW (eds) The social biology of wasps. Cornell University Press, Ithaca, NY

Röseler PF, Röseler I, Strambi A (1980) The activity of corpora allata in dominant and subordinated females of the wasp Polistes gallicus. Insect Soc 27:97–107

Saito F, Nguyen LPT, Carpenter JM, Kojima J (2006) A new *Eustenogaster* species (Hymenoptera: Vespidae; Stenogastrinae), the first hover wasp known to overwinter on the nest. Am Mus Novit 3534:1–12

Saito F, Nguyen LPT, Kojima J (2009) Colony cycle of a "temperate" hover wasp, *Eustenogaster nigra*, with special reference to overwintering of males in an enveloped nest together with virgin females (Hymenoptera, Vespidae, Stenogastrinae). Insect Soc 56:49–54

Sakagami SF, Yamane S (1990) A behavior inventory of the females of two stenogastrine wasps *Parischnogaster mellyi* and *Liostenogaster vechti* (Hymenoptera, Vespidae). In: Sakagami SF, Ohgushi R, Roubik DW (eds) Natural history of social wasps and bees in equatorial Sumatra. Hokkaido University Press, Sapporo

Samuel C (1987) Factors affecting colony size in the stenogastrine wasp *Liostenogaster flavolineata*. PhD thesis, University of Malaya, Kuala Lumpur
Shreeves GE, Field JP (2002) Group size and direct fitness in social queues. Am Naturalist 159:81–95
Spradbery JP (1975) The biology of *Stenogaster concinna* Van der Vecht with comments on the phylogeny of Stenogastrinae (Hymenoptera: Vespidae). J Aust Entomol Soc 14:309–318
Spradbery JP (1989) The nesting of *Anischnogaster iridipennis* (Smith) (Hymenoptera: Vespidae) in New Guinea. J Aust Entomol Soc 28:225–228
Stacey PB, Ligon JD (1991) The benefits of philopatry hypothesis for the evolution of cooperative breeding: habitat variance and group size effects. Am Nat 137:831–846
Strassmann JE, Meyer DC (1983) Gerontocracy in the social wasp, *Polistes exclamans*. Anim Behav 31:421–438
Strassmann JE, Queller DC (1989) Ecological determinants of social evolution. In: Breed MD, Page RE (eds) The genetics of social evolution. Westview, Boulder
Strassmann J, Hughes CR, Turillazzi S, Solis CR, Queller DC (1994) Genetic relatedness and incipient eusociality in stenogastrine wasps. Anim Behav 48:813–821
Sumner S (1999) Conflicts over reproduction in facultatively eusocial hover wasps. PhD thesis, University College London, UK
Sumner S, Field J (2001) Highly polymorphic microsatellite loci in the facultatively eusocial hover wasp, *Liostenogaster flavolineata* and cross-species amplification. Mol Ecol Notes 1:229–322
Sumner S, Casiraghi M, Foster W, Field J (2002) High reproductive skew in tropical hover wasps. Proc R Soc Lond B 269:179–186
Tibbetts EA (2002) Visual signals of individual identity in the wasp *Polistes fuscatus*. Proc R Soc Lond B 269:1423–1428
Tibbetts EA, Dale J (2004) A socially enforced signal of quality in a paper wasp. Nature 432:218–222
Tibbetts EA, Mettler A, Levy S (2010) Mutual assessment via visual status signals in *Polistes dominulus* wasps. Biol Lett 6:10–13
Turillazzi S (1985a) Colonial cycle of *Parischnogaster nigricans serrei* (Du Buysson) in West Java (Hym. Stenogastrinae). Insect Soc 32:43–60
Turillazzi S (1985b) Associative nest foundation in the wasp: *Parischnogaster alternata*. Naturwissenschaften 72:100–102
Turillazzi S (1986) Colony composition and social behaviour of *Parischnogaster alternata* Sakagami (Hymenoptera Stenogastrinae). Monit Zool Ital 20:333–347
Turillazzi S (1988) Social biology of *Parischnogaster jacobsoni* (Du Buysson) (Hymenoptera, Stenogastrinae). Insect Soc 35:133–143
Turillazzi S (1989) The origin and evolution of social life in the Stenogastrinae (Hymenoptera Vespidae). J Insect Behav 2:649–661
Turillazzi S (1990a) Social biology of *Liostenogaster vechti* Turillazzi 1988 (Hymenoptera Stenogastrinae). Trop Zool 3:68–87
Turillazzi S (1990b) Notes on the biology, social behaviour and nest architecture of *Metischnogaster drewseni* (Saussure) (Hymenoptera Stenogastrinae). Boll Zool 57:331–339
Turillazzi S (1990c) Socialità ed architettura del nido nelle vespe Stenogastrinae (Hymenoptera, Vespidae). Atti Acc Naz Ital Entomol Rendiconti XXXVIII:1–24
Turillazzi S, Carfi S (1996) Adults and nest of *Liostenogaster pardii* n.sp. (Hymenoptera Stenogastrinae). Trop Zool 9:19–30
Turillazzi S, Francescato E (1989) Observation on the behaviour of male stenogastrine wasps (Hymenoptera, Vespidae, Stenogastrinae). Actes Coll Insect Soc 5:181–187
Turillazzi S, Gerace L (1992) Size and characteristics of colonies of *Eustenogaster fraterna* (Hymenoptera Vespidae Stenogastrinae). Ethol Ecol Evol (Special Issue) 2:65–67
Turillazzi S, Hansell MH (1991) Biology and social behaviour of three species of *Anischnogaster* (Vespidae, Stenogastrinae) in Papua New Guinea. Insect Soc 38:423–437

Turillazzi S, Pardi L (1982) Social behavior of *Parischnogaster nigricans serrei* (Du Buysson) (Hymenoptera Stenogastrinae) in Java. Ann Entomol Soc Am 75:657–664

Turillazzi S, Marino Piccioli MT, Hervatin L, Pardi L (1982) Reproductive capacity of single foundress and associate foundress females of *P. gallicus* (Hymenoptera Vespidae). Monit Zool Ital 16:75–88

Turillazzi S, Cervo R, Cavallari I (1990) Invasion of the nest of *Polistes dominulus* by the social parasite *Sulcopolistes sulcifer* (Hymenoptera, Vespidae). Ethology 84:47–59

Turillazzi S, Cervo R, Dani FR (1997) Intra- and inter-specific relationships in a cluster of Stenogastrine wasp colonies (Hymenoptera; Vespidae). Ethol Ecol Evol 9:385–395

van der Vecht J (1977) Studies of oriental Stenogastrinae (Hymenoptera: Vespoidea). Tijdschr Entomol 120:55–75

West-Eberhard MJ (1978) Polygyny and the evolution of social behavior in wasps. J Kansas Entomol Soc 51:832–856

Williams FX (1928) Studies in tropical wasps their hosts and associates (with descriptions of new species). Bull Exp Sta Hawaii Sugar Planters' Assoc Entomol 19:1–179

Yamane S, Sakagami S, Ohgushi R (1983) Multiple behavioral options in a primitively social wasp, *Parischnogaster mellyi*. Insect Soc 30:412–415

Yamane S, Sakagami S, Ohgushi R (1990) Social behavior in the stenogastrine wasp *Parischnogaster mellyi* (Hymenoptera Vespidae). In: Sakagami SF, Ohgushi R, Roubik DW (eds) Natural history of social wasps and bees in equatorial Sumatra. Hokkaido University Press, Sapporo

Yoshikawa K, Ohgushi R, Sakagami SF (1969) Preliminary report on entomology of the Osaka City University 5th Scientific Expedition to Southeast Asia 1966. With description of two new genera of stenogasterine [sic] wasps by J. van der Vecht. Nat Lif SEA 6:153–182

Zanette LRS, Field J (2009) Cues, concessions and inheritance: dominance hierarchies in the paper wasp *Polistes dominulus*. Behav Ecol 20:773–780

Chapter 5
Social Communication

5.1 Communication in Social Insects

A high flow of information between the components of a group is the main requisite for the formation of stable and highly organised social structures. In a colony of social insects, individual members must be coordinated in their behaviours in order to ensure the best responses to various environmental stimuli, such as attack by a predator or exploitation of a rich food source before it is used by individuals from other colonies (Detrain et al. 1999). Efficient communication mechanisms are thus the basis for efficient social organisation.

Social complexity therefore relies on communication which occurs when the behaviour of a receiver is influenced by a message produced by a releaser individual (Krebs and Davies 1987). Communicative pathways are the results of an evolutionary process and the kinds of messages can be countless, transferred along four or five main channels. Signals are specific messages which are the result of a co-adaptive process between releasers and receivers, cues, on the other hand can transmit information but have not been moulded by natural selection in the course of a co-adaptation process (Seeley 1989).

5.2 Channels for Information Transfer

The communication channels used by social insects are mainly four: visual, tactile (contact of body parts), vibrational and chemical. Social bees may also use an acoustical one, capable as they are of perceiving air vibrations (Dreller and Kirschner 1993). In social wasps all these channels are used by various species and in different contexts but hover wasps seem to present a quite limited set of possible kinds of message with respect to those of more socially evolved species, mainly using chemical and visual channels. Any kind of signal or cue needs the existence of anatomical and/or morphological structures for emission and

S. Turillazzi, *The Biology of Hover Wasps*, DOI 10.1007/978-3-642-32680-6_5,

anatomical and/or morphological structures suitable for reception. Visual signals, for example, are perceived by the eyes but can consist in body colourations, behavioural displays, etc. and require that the environment is illuminated to some extent (unless the signal itself consists of flashes of lights as in fireflies). Moreover, visual signals can only be fully perceived by a receiver if the signaller is not covered by obstacles, which could even be other possible receivers. The energy to generate visual signals is usually limited in the case, for example, of evident body colourations, but can increase if they consist of elaborate behavioural displays or by colourations which request high physiological costs or tradeoffs to be built or maintained. In any case the passage of information through this channel is almost instantaneous.

Vibratory signals need a substrate to be transmitted and in the case of social wasps this is usually represented by the nest itself. The success of transmission is determined by the material and the structure of the substrate. Wax, for example, is not a very good material for transmitting vibrations, but it is much better when, at the right temperature, it forms the complex net of cell borders which assures communication between individuals during the waggle dance of honeybees (Nieh and Tautz 2000). In general, producing vibratory signals of a certain strength implies a relevant quantity of energy.

Chemical signals are relatively cheap to produce and can be used when it is necessary for the information to reach a high number of receivers, even in the absence of light. Signals can be maintained for a relatively long time. They are only perceived when the receptors of the receiver enter into contact with the message substances themselves; however, if they consist of volatile substances the passage of information is influenced by distance and air turbulences (Bruschini et al. 2010).

5.3 Social Communication in Hover Wasps

I choose here to treat communication in hover wasps by analysing the different contexts in which certain kinds of information transfer occur.

5.3.1 Adults–Immature Brood Interactions

Tactile messages are evidently used by adult individuals in interactions between themselves and large size larvae. Larvae react to the stimuli of adults biting them by opening themselves up like sphincters to receive solid or liquid food or to permit the adults to take some still uneaten food pellets. At present it is unknown if vibratory signals of the kind used by *Polistes* or by vespine wasps, such as abdominal wagging or antennal drumming (Gamboa and Dew 1981, Pratte and Jeanne 1984, Brennan 2007, Bernard and Brennan 2007), to communicate with larvae are used by some hover wasps; however, no particular behaviour which could be interpreted as a way of producing vibrations in the context of larval rearing has ever been described in the species studied so far. However, in *E. fraterna* Francescato et al. (2002a) reported one particular body vibration, consisting in a longitudinal body oscillation lasting 2–3 s, performed by adult wasps on different immature stages (including eggs and pupae), the function of which has not yet been established.

5.3.2 Adult–Adult Interactions

5.3.2.1 Recognition

The ability of members of a social group to distinguish individuals belonging to their own colony from aliens is critical in preventing non-colony members from exploiting the resources and worker force of the colony (Starks 2004). The recognition system consists of three main components: the expression component, the perception component and the action component and involves two actors which are termed "cue-bearer" and "evaluator" (Starks 2004). The expression component refers to the nature and production of the cues by the cue-bearer, which is used for the process of recognition. The perception component comprises the mechanisms which renders cue perception possible, the elaboration of an internal template and comparison of this with all the possible phenotypes encountered by the evaluator; the action component is represented by the particular behavioural or physiological reaction of the evaluator following recognition (Reeve 1989, Gamboa 1996, 2004).

Nestmate Recognition

Nestmate recognition seems to be widespread in hover wasps. The first systematic studies on this matter were led by Cervo et al. (1996) who performed a series of behavioural experiments and observations on *Parischnogaster jacobsoni* and on two species of the genus *Liostenogaster*, *L. vechti* and *L. flavolineata*. However, observations by previous authors on the aggressive reactions of resident females to alien females had already been recorded in various species (Samuel 1987). We observed that resident individuals of colonies of *P. jacobsoni*, which build isolated nests consisting in a long series of cells placed along thin thread-like substrata, usually attack conspecifics approaching the colony or aggressively chase any alien individuals landing on the nest away. We presented the colony residents with immobilised individuals taken from the same or different colonies and observed that nestmates were accepted peacefully (only inspected with antennae) while non-nestmates were fiercely attacked with bites and abdomen slaps. The accuracy with which the individuals of this species discriminated nestmates from aliens (only 1 out of 14 colonies—7 %—showed peaceful behaviour towards alien females) was actually noticeably higher that that shown by *L. vechti* and *L. flavolineata* which seemed to accept a fairly high percentage (24 % and 30 % respectively) of alien females on their nests without any apparent aggressive behaviours. This fact was explained as a kind of erroneous recognition due to the high rate of shifting from one nest to another in the overcrowded clusters formed by the colonies of these two species (also observed by Samuel 1987, in *L. flavolineata*) (Cervo et al. 1996, 2002). But in any case the presentation experiments demonstrated that all three species had the capacity to discriminate between nestmates and alien individuals.

Nestmate recognition was also studied in *Parischnogaster striatula*, another species which builds isolated nests with a spiral architecture. In this species we demonstrated for the first time that, as occurs in other social insects, cuticular hydrocarbons are the cues which the wasps use to recognise their nestmates. Zanetti and co-workers, after ascertaining that members of the same colony were peacefully welcomed while members of other colonies were aggressively treated by the residents of one nest, captured and killed by freezing a number of individuals from the same and other colonies, washed them for 5 h in hexane (to obtain extracts of cuticular lipids), then for a further 10 h (in order to eliminate any possible apolar substance from their bodies), and finally presented these lures to the residents. At this point the lures were treated peacefully, but when the nestmate lures were re-applied with the extracts from the alien females, they were attacked, while alien lures treated with the extract from nestmate females were peacefully welcomed. The results of these experiments strongly support the involvement of cuticular lipids in nestmate recognition in this species (Zanetti et al. 2001).

Similar experiments were carried out on *L. flavolineata* by Cervo et al. (2000b, 2002) who also performed bioassays to test the involvement of Dufour's gland secretion in the recognition process. The behavioural reactions of the colonies indicated that chemical cues on the body surface are necessary for nestmate discrimination. The importance of the relative quantity of cuticular linear and branched alkanes and mono un-saturated alkenes were confirmed also in this species.

All these behavioural investigations were accompanied by chemical analyses and correlational statistical studies. In general, the cuticular chemical profile of each individual was established using Gas Chromatographic analysis and all the major peaks were identified by means of their spectra obtained with a Mass Spectrometer coupled with the Gas Chromatograph (Fig. 5.1). The areas of the peaks of the chromatogram were treated as independent variables and the values of the various variables were processed with a multivariate statistical analysis. Different authors from my group analysed the cuticular chemical profiles of the females of various species of hover wasps (Zanetti et al. 2001 for *P. striatula*, Cervo et al. 2002 for *L. flavolineata*, Beani et al. 2002, Destri et al. 2002 and Turillazzi et al. 2008 for male and female *P. mellyi* and Turillazzi et al. 2004 for *E. fraterna* and *L. vechti*).

Alkenes are the principal components of the cuticular mixture in all the species, whereas this class of compounds is not so frequent in *Polistes* wasps (Espelie et al. 1990; Singer et al. 1998, Sledge et al. 2001). On the contrary, methyl alkanes, so numerous and abundant in *Polistes*, are only found in *Liostenogaster vechti* as

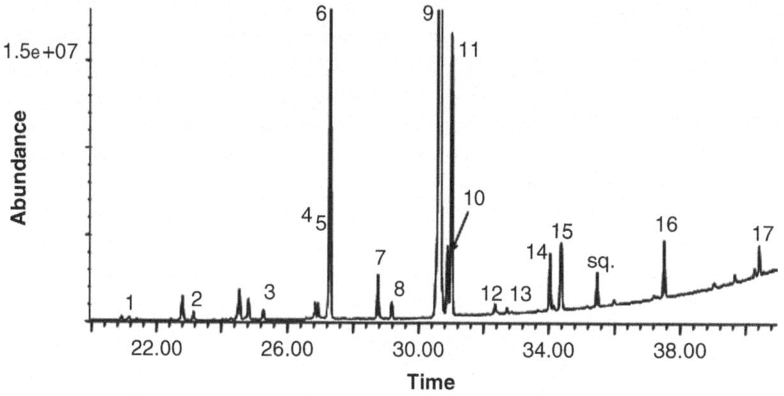

Fig. 5.1 Gas Chromatographic cuticular profile of a female *Eustenogaster fraterna*. Numbered peaks correspond to those reported in Table 5.1 for the same species

monomethyl alkanes and in *P. mellyi* as mono and di-methyl alkanes; aliphatic alcohols have been found in *E. fraterna*, *L. flavolineata* and *P. striatula* (see Table 5.1). Differences, both quantitative and qualitative, between species of the same genus can be relevant, especially in *Parischnogaster* and *Liostenogaster*.

Correlational analyses showed that in *P. striatula* (Zanetti et al. 2001), *L. flavolineata* (Cervo et al. 2002) and *P. mellyi* (Turillazzi et al. 2008) the chemical profiles of individuals belonging to the same colonies are statistically more similar to each other than those from individuals belonging to other colonies. Similar analyses were performed for *E. fraterna* by Matthew Sledge (unpublished) (Fig. 5.2).

However Breed (1998) observes how differences and similarities in the cues borne by some individuals do not mean that such cues are in fact perceived and used in the recognition process. In other social insects supplementation experiments, consisting in the addition of single synthetic chemical compounds on the body of focal animals or dead lures to check if recognition by other members of the group was altered or not, proved to be important for furnishing further evidence on the importance of cuticular hydrocarbons in nestmate recognition (Meskali et al. 1995, Breed et al. 2004, Dani et al. 2001, Dani et al. 2005, Ruther et al. 2002). Breed himself, however, observes that an appropriate experimental design to test the effect of single compounds on recognition without altering the concentration of all the others has not yet been settled on (Breed 1998).

We worked on the cuticular chemistry and nestmate recognition of another common species of *Parischnogaster*, *P. mellyi*, which is somewhat particular because its cuticular chemical profile is mainly composed of methyl-branched alkanes but very few alkenes (Destri et al. 2002). After ascertaining that residents discriminate between alien and nestmate females, we performed an experiment in which we applied an alkane and a methyl branched alkane with the same number of C atoms in their molecules (nonacosane—C_{29}—and methyl-nonacosane, me-C_{29})

Table 5.1 List of chemical compounds detected on the cuticle of females of *L. flavolineata* (Cervo et al. 2002), *P. striatula* (Zanetti et al. 2001) and *E. fraterna* (M. Sledge unpublished)

See Fig. 5.1	Compound	*Liostenogaster flavolineata* Average relative percentage (SD) ($N = 41$)	*Parischnogaster striatula* Average relative percentage (SD), ($N = 44$)	*Eustenogaster fraterna* Average relative percentage (SD), ($N = 28$)
1	*n*-octadecane (*n*-C18)	–	–	0.328 (0.428)
2	*n*-nonadecane (*n*-C19)	–	–	0.635 (0.526)
3	*n*-icosane (*n*-C20)	–	–	0.592 (0.444)
4	9-heneicosene (9-$C_{21:1}$)	0,142 (0.141)	0.117 (0.121)	0.466 (0.326)
5	7-heneicosene (7-$C_{21:1}$)	0.058 (0.027)	1.021 (1.156)	0.634 (0.984)
	1-octadecanol (1-C_{18OH})	0.168 (0.100)	0.083 (0.184)	–
6	*n*-heneicosane (*n*-C_{21})	10.637 (2.968)	15.005 (4.577)	14.374 (1.773)
7	9-docosene (9-$C_{22:1}$)	0.861 (0.374)	0.885 (1.774)	1.556 (0.322)
	7-docosene (7-$C_{22:1}$)	0.085 (0.193)	0.473 (0.487)	–
8	*n*-docosane (*n*-C_{22})	0.592 (0.162)	0.932 (0.291)	0.681 (0.300)
	6,9-tricosadiene (6,9-$C_{23:2}$)	0.343 (0.548)	0.359 (0.232)	–
9	9-tricosene (9-$C_{23:1}$)	63.381 (8.799)	33.686 (18.367)	58.493 (6.563)
	7-tricosene (7-$C_{23:1}$)	0.589 (0.371)	15.625 (15.551)	–
10	1-heicosanol (1-C_{20OH})	0.070 (0.039)	0.364 (0.441)	2.062 (2.910)
11	*n*-tricosane (*n*-C_{23})	12.659 (6.196)	18.275 (9.704)	10.927 (3.577)
	11-methyltricosane (11-MeC_{23})	0.280 (0.133)	–	–
12	9-tetracosene (9-$C_{24:1}$)	0.720 (0.229)	0.271 (0.124)	0.378 (0.436)
	8-tetracosene (8-$C_{24:1}$)	–	0.081 (0.178)	–
	7-tetracosene (7-$C_{24:1}$)	–	0.072 (0.134)	–
13	*n*-tetracosane (*n*-C_{24})	0.170 (0.103)	0.228 (0.662)	0.221 (0.274)
	6,9-pentadiene (6.9-$C_{25:2}$)	0.068 (0.034)	–	–
14	9-pentacosene (9-$C_{25:1}$)	5.002 (3.528)	2.961 (2.163)	2.055 (1.194)

(continued)

Table 5.1 (continued)

See Fig. 5.1	Compound	*Liostenogaster flavolineata* Average relative percentage (SD) ($N = 41$)	*Parischnogaster striatula* Average relative percentage (SD), ($N = 44$)	*Eustenogaster fraterna* Average relative percentage (SD), ($N = 28$)
	7-pentacosene (7-$C_{25:1}$)	0.120 (0.064)	0.537 (0.458)	–
	1-docosanol (1-C_{22OH})	0.014 (0.033)	0.072 (0.327)	–
15	*n*-pentacosane (*n*-C_{25})	2.785 (1.886)	1.361 (0.958)	3.124 (1.803)
	n-hexacosane (*n*-C_{26})	0.046 (0.041)	0.306 (0.446)	–
	9-heptacosene (9-$C_{27:1}$)	0.238 (0.238)	0.502 (0.367)	–
	7-heptacosene (7-$C_{27:1}$)	0.010 (0.013)	0.073 (0.100)	–
16	*n*-heptacosane (*n*-C_{27})	0.795 (0.584)	4.179 (3.087)	2.363 (1.160)
	n-octacosane (*n*-C_{28})	0.036 (0.094)	0.166 (0.162)	–
	9-nonacosene (9-$C_{29:1}$)	0.033 (0.037)	0.275 (0.324)	–
	17-hexacosen-1-ol (17-$C_{26:1OH}$)	–	0.291 (0.523)	–
17	*n*-nonacosane (*n*-C_{29})	0.097 (0.083)	–	1.101 (0.881)

For each compound the average relative percentage and the Standard Deviation are reported. Numbered compounds are those labelled in Fig. 5.1 for *E. fraterna*

in a 20 mg/ml hexane solution, on live, but immobilised, females taken and presented at a short distance to the residents of their own colonies. We also tested an alkane (tricosane—C_{23}) and its unsaturated equivalent, the alkene (Z)-9-tricosene ((Z)-9-$C_{23:1}$)) which were applied with the same technique on other individuals. The first set of supplementation experiments, performed on six colonies, showed that the wasps treated with the methyl-branched compound elicited only aggression from their resident nestmates and significantly more aggression than the ones treated with the linear alkane equivalent. Instead, in the second set of experiments, held on nine colonies, we observed no significant difference in the reaction to saturated and unsaturated compounds, while aggressive and peaceful reaction to the single compounds were nearly similar. We concluded that nestmate recognition in this species could be based exclusively on the perception of hydrocarbons possessing particular functional groups, similarly to what is hypothesised for other social insects (Breed and Julian 1992, Vander Meer and Morel 1998, Dani et al. 2001, Destri et al. 2002.)

Regarding the "development" of the chemical template which is used for the recognition process by the evaluator individuals, as yet we have no strong evidence

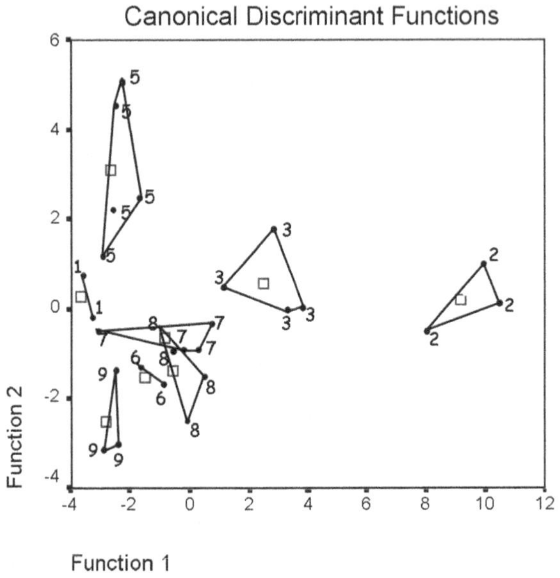

Fig. 5.2 Plot of the first and second discriminant components of the Discriminant Analysis based on hydrocarbons proportions present on the cuticle of 22 females of *Eustenogaster fraterna* belonging to 8 different colonies Groups of individuals belonging to the same colony (from 1 to 9—colony 4 is missing) are numbered and grouped with the indication of relative group centroid (*white square*). (By M. Sledge, unpublished)

regarding the Stenogastrinae. We know that in *Polistes* this is obtained by young imagoes, just after emergence, from the material of the maternal nest which is covered with the cuticular hydrocarbons which the adults, especially the more dominant ones, continuously leave on the surface (Gamboa 1996, 2004, Dani et al. 1992). In hover wasps we found that hydrocarbons similar to the cuticular ones can also be found in the secretion of the Dufour's gland and in the secretion found on the eggs and little larvae (Sledge et al. 2000) and even in the gut excretions of males and females (Beani et al. 2002). The nests of some species are built with vegetal matters and in this case we could have a situation similar to that of *Polistes*, but in various species of *Liostenogaster* the nest is built with mud and this could be a drawback for the application of adult hydrocarbons. We studied this problem in *P. mellyi* where we tested the possibility that the secretion placed on the eggs could be the cue for the elaboration of the chemical template distinguishing a colony. A chemical correlational study furnished evidence that it was possible to correctly discriminate 100 % of the members of five different colonies with a Discriminant Stepwise Analysis performed on their cuticular chemical profiles and 86.7 % of them with their Dufour's gland compounds. Similarly it was also possible to discriminate correctly 94 % of the eggs found in the different nests from the secretion on them. However, behavioural observations performed after exchange of secretion from egg to egg from different colonies showed that females did not discriminate between their own and alien eggs. These experiments cast some doubts on the possibility that the Dufour's gland secretion actually does represent the basis for the formation of the template even if the scarcity of data leaves the problem still open (Turillazzi et al. 2008).

Visual cues could be another way of recognising alien individuals from nestmates. The very first discrimination that residents of a colony could use could well be the way their conspecifics fly as they approach the nest. Zanetti et al. (2001) already observed that females of *Parischnogaster striatula* leave their nests to attack conspecifics in flight, as well as other insects and artificially moved objects which perform a zig-zag hovering flight close to the nest, but they do not attack conspecifics which land directly on the nest. The authors also discovered that the members of the colony use a direct approach to the nest while aliens hover in front of the nest before landing and also found that males approach alien nests directly more often than females. These data confirm the observations by M.J. West-Eberhard (1969) who noted that the direct or hesitant way in which conspecifics approach the nests in *Polistes fuscatus* could determine the first reaction by resident females towards landing individuals. The same phenomenon was observed by Francescato et al. (2002a) in *E. fraterna* where females guarding the tubular entrance of their enveloped nests react with abdomen beatings when alien females fly with uncertainty around the nest.

A more accurate visual recognition would be possible particularly in those species which present evident colour markings on their head and face. Facial colouration patterns are particularly manifested not only in species of *Liostenogaster* but also in *Eustenogaster* and some species of *Parischnogaster*. *Liostenogaster vechti* and *L. flavolineata* have quite enlarged and distinctive brown spots on a yellowish background that can be easily appreciated by a human observer and even used for distinguishing one individual from another (Fig. 5.3).

A recent research performed on colonies of these two species showed that in *L. flavolineata* the possibility does exist that resident females can discriminate alien from nestmate individuals on the basis of their facial markings, as similarly occurs in *Polistes fuscatus* (Tibbetts 2002). Females with brown markings artificially altered with paint were aggressively attacked when presented to their own nestmates in contrast to females whose markings were painted but not changed (Baracchi et al. 2012).

Immature Brood and Nest Recognition

The ability of social wasps to recognise immature brood and the nest itself have been studied by various researchers, mainly using nest exchange experiments. This is a technique in which all the adults are chased away from two nests which are then quickly swapped over in position, then waiting to observe the behaviour of individuals returning to the right position but wrong nest. Klahn and Gamboa (1983) were the first to use these kinds of experiment in *Polistes* to check if foundresses could recognise their own brood. The same technique was also used by Cervo and Turillazzi (1989) to assess nest recognition capacity in foundresses of

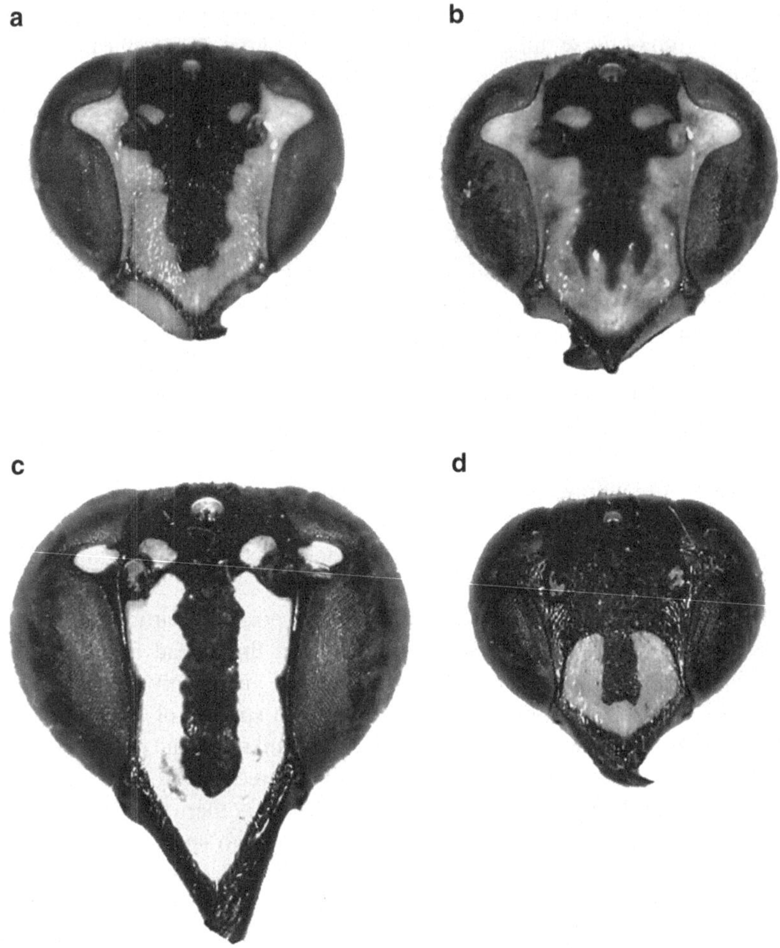

Fig. 5.3 Facial markings in females of (**a**) *L. vechti*, (**b**) *Liostenogaster flavolineata*, (**c**) *E. calyptodoma* and (**d**) *Parischnogaster alternata*

Polistes gallicus. In hover wasps, females of *Parischnogaster jacobsoni* proved capable of recognising alien immature brood in switched nests, killing and eliminating much of the eggs and younger larvae with respect to control broods present in manipulated but not switched nests (Cervo et al. 1996). The same occurred in switched nests of *P. mellyi* (Turillazzi et al. 2008). It is important to note that in hover wasps only the youngest immature brood are eliminated while more mature larvae and pupae are left intact and reared by the foster wasps; in *Polistes*, on the contrary, the most developed brood is eliminated and eggs and small larvae are usually left intact (Cervo and Turillazzi 1989). When foundress hover wasps land on the switched, alien nest, they show a high degree of agitation performing a high number of cell inspections and patrolling the nest on its surface and in flight, but we never observed particular behaviours of the kind performed by *Polistes* wasps.[1] The latter, in fact, begin to drag their abdomens over the nest surface probably applying their own chemical signature (CHCs produced by the van der Vecht organ) (Cervo and Turillazzi 1989). The absence of behaviours like this in the species of Stenogastrinae studied so far, casts some doubts on the utilisation of chemical cues for the recognition of the nest and immature brood. In *Polistes,* however, this behaviour seems to be aimed at spreading the cues on the nest paper that will be used for the formation of a colony recognition template by the newly emerged brood rather than for changing the cues for recognising the brood itself.

In hover wasps, as I stressed above, any colony component examined (including secretion deposited on eggs and nest material) has the potential chemical information to act as a cue for the recognition of the colony (Turillazzi et al. 2008). Nest exchange experiments performed on colonies of *P. mellyi* showed that the average number of eggs and larvae which disappeared from exchanged nests was significantly higher than that observed on nests simply detached and then repositioned in the same place, even only in the very first 48 h after switching. Other experiments, consisting in the exchange of secretion between eggs of different nests or of nest materials, failed to throw in evidence discriminatory behaviours of the resident females of the tested colonies (Turillazzi et al. 2008). Visual recognition of nests, on the basis of different shapes or cell arrangement or of the different disposition of immature brood, can also probably be excluded due to the fact that in their natural environment the nests are often exposed to significant changes provoked by meteorological conditions, partial destruction by predators like *Vespa tropica* or by conspecifics stealing nest material.

Thus, at present, we can say that *P. mellyi* is capable of limited immature brood recognition but the problem of ascertaining which cues make the recognition process of immature brood possible and the formation of the recognition template remains open to future investigation.

Dominance–Subordinance

We recently tested the importance of the dark brown colouration that female *L. vechti* and *L. flavolineata* display on the face. This dark spot varies considerable in shape and margins so that all the individuals can be readily recognised even by a human observer (see Fig. 5.3).

As explained in the previous chapter we found significant correlations between the size of the brown colouration and ovarian development and rank occupied in the dominance hierarchy of females belonging to colonies of *L. vechti*. In general, the wider the relative dark colouration, the higher the rank in the hierarchy occupied by

---

[1] Recently, however, I observed a female *Parischnogaster mellyi* repeatedly pressing the tip of her partly everted anus over the surface of the nest after repelling the attacks of a possible intruder: this manoeuvre suggested a probable nest marking.

that wasp. Thus, in this species the facial markings would function as a status badge but one which, in any case, must be tested by other nestmates before it is matched with the effective worthiness of the bearer individual (Baracchi et al. 2012). A badge of status can be useful as a visual signal also in the complex game which is part of the shifting of individuals between different colonies, especially in nest clustering species; in fact floater females (usually relatively young individuals) could test the possibility of joining new colonies by hovering in front of their nests or landing briefly to challenge residents before possibly adventuring into gruelling fights.

In *L. flavolineata* we did not find the same correlation as in *L. vechti* but experiments performed by my team on field colonies showed that nest residents react aggressively towards nestmate females with artificially altered brown spots, suggesting that in this species facial marking could be used as individual recognition cues (see Tibbetts 2002). If this is the case, single individuals could learn not only that another individual is a member of the same colony but also its position in the hierarchical queue. This interpretation could also explain the mechanism which would enable an individual female to assess the relative positions in the gerontocratic hierarchy in a *L. flavolineata* colony (Bridge and Field 2007, Field 2008).

All the behavioural patterns described in the interactions observed between different individuals in a colony could be included in tactile communication, beginning with head butting and ending with the somewhat spectacular antennal fencing described in *L. flavolineata* (Samuel 1987) and *L. vechti* (Turillazzi 1990, Sakagami et al. 1990) (see Chap. 3). In *E. calyptodoma* Hansell (1987a) observed that solicitation of food was preceded by gentle antennal strokes of the begging individuals on the antennae of the food bearer. Abdomen bending is instead a behaviour which could function as a visual signal showing the predisposition of the individual to react aggressively to a competing partner; as we have seen (Chap. 3) this posture is also used in the anti-predatory behaviour. There is no evidence that during these behaviours chemical signals are also produced by the wasps.

We can probably also include in the dominance–subordinance section the particular behaviour observed in species of the genus *Eustenogaster*. I refer to the sound (or drumming) provoked by guarding females when conspecifics (or predators) approach or land on their nests. The behaviour has been described by Mike Hansell in *E. calyptodoma* who observed that resident guarding females (usually the eldest in the colony) react by vibrating their gasters rapidly up and down against the nest envelope producing a loud noise (Hansell 1987). Also alien, and potentially usurper females, which try to enter the defended nest perform vibratory movements but produce much less noise as they are not inside the entrance tube far enough to strike the nest envelope effectively (Hansell 1987). This particular behaviour has been reported in other congeneric species: *E. eximia* by Krombein (1991), *E. fraterna* and *E. sp* (prob. *micans*) by Francescato et al. (2002a). We studied this particular phenomenon more deeply by performing observations and experiments in some species and concluding that the sound probably functions as an acoustical signal made with the aim of warning conspecifics and small predators (ants) that the nest is engaged and discouraging any attempt of usurpation and predation. In particular we observed the reactions of resident females and the occurrence of beating behaviour while presenting dead or live alien or nestmate females and small predators (ants or hornets) at the nest entrance and recording the sounds of the beats with a stylus laid on the top of the nest envelope. Beating occurred in 7 out of 10 cases in which alien females approached the nest with a zigzag flight and landed on the nest before trying to enter, but not when nestmates returned directly in flight to the nest. With the presentation experiments, we obtained reactions from the resident females when we presented alien or nestmate females at the entrance of the nest or placed live ants on the nest envelope. In no case did the residents come to the entrance spout of the nest to antennate the presented lures, which proves that they can see outside through the entrance or through the large number of holes present in the lower part of the envelope (Lunghi 1999, Francescato et al. 2002b).

In another experiment we chased the resident females of 14 colonies off their nests and, on their return, we played back the recorded drumming noise with a mini loudspeaker placed near the nests. Only in 3 out of 14 cases did the females avoid entering the nest, showing no significant differences with control experiments (where no noise was played back and all the wasps entered their nests), yet they showed a significant delay in entering the nest with respect to controls. Of course the results of these experiments could have been influenced by the fact that the tested females were actually the owners of the nests and they had more drive to enter (Lunghi 1999, with Francescato and myself).

Playback experiments open the question of whether female *Eustenogaster* are able to perceive airborne sounds; actually we could not completely reject the hypothesis that they perceived the vibrations of the nest entrance spout shaken by air vibrations emitted by the loudspeaker. It is in any case peculiar that this sort of signal is present in all the species of *Eustenogaster* but works quite well in *E. calyptodoma* where the nest envelope is particularly large and completely independent from the cell comb (see also Chap. 6, Fig. 6.52). It is also curious that the nest envelope is quite similar to an acoustic (Helmholtz) resonator, an instrument that at a given frequency amplifies the signal intensity, and sound is clearly audible even some metres distance (Francescato et al. 2002b). To find some correlates with the intensity of the sound produced by a beating wasp, I and some students of mine (D. Lambardi, I. Ortolani, D. Baracchi and M. Zaccaroni) recorded the drumming (Fig. 5.4) of various females in nests of different size but we were not able to find any relation between the main sound parameters (intensity, frequency, etc.) and certain colony characteristics (such as the number and kind of immature brood present or the size of the nest or of the resident females). We were only able to ascertain that the larger the volume of the envelope of a nest, the higher the intensity of sound produced by a small glass ball falling on it from a fixed height, even if the relation is not statistically significant ($P = 0.084$). Our, not yet discarded, hypothesis is that the particular and quite enlarged envelope of this species could have evolved not only under the pressure of parasitoids but also as a system to amplify the output of beating behaviour in nest defence against

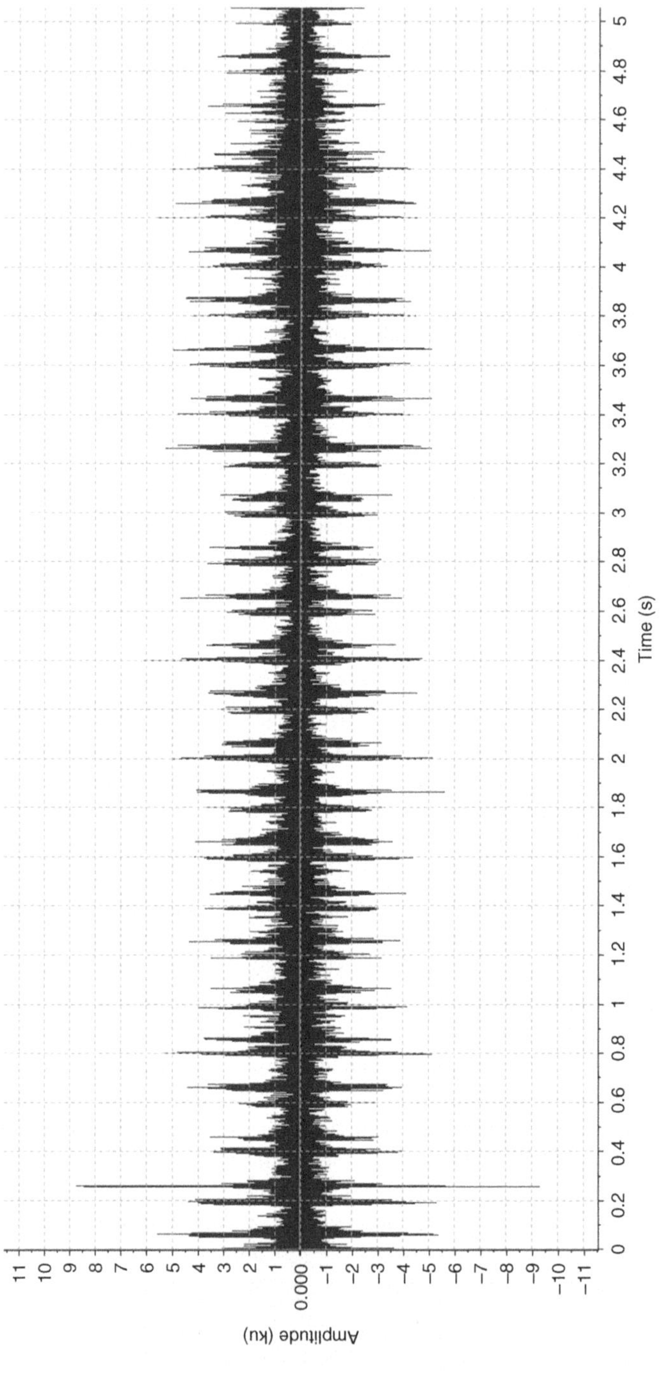

Fig. 5.4 Oscillogram of the beating produced by a female *Eustenogaster calyptodoma* drumming her abdomen against the interior of the nest envelope during the attack of a usurping alien female (Lambardi D., Ortolani I., Baracchi D., Zaccaroni M., Turillazzi S., unpublished)

conspecific usurpation. The question remains opens and awaits a solution probably employing more accurate measurements and opportune experiments.

Fertility

In social insects cuticular lipids are involved in many communicative functions, one of the most important in intra-colony communication is signalling reproductive status by fertile individuals: egg-laying individuals are predicted to produce signals of their reproductive potential that are used by their nestmates to perform or adjust their helping behaviour in order to maximise their inclusive fitness (Keller and Nonacs 1993). This has been demonstrated in several species of ants, bees and wasps (Peeters et al. 1999, Liebig et al. 2000, Ayasse et al. 1995, Sledge et al. 2001) where variations in the cuticular hydrocarbon chemical profile may give information on the reproductive capacities of the individuals. In hover wasps the lipid cuticular profiles of females of four species belonging to three genera (*Liostenogaster*, *Parischnogaster* and *Eustenogaster*) were analysed in relation to the degree of their ovarian development and relative to the social situation of their own colony. In all the species examined, individuals with developed ovaries differed in the abundance of one or more cuticular compounds from those with undeveloped ovaries. Moreover, the relative quantity of one or more compounds was correlated with the degree of ovarian development. However, Stepwise Discriminant Analysis performed on chemical substances correctly identified more than 95% of the fertile and non fertile individuals in *P. striatula* but only 65% of the individuals in the two *Liostenogaster* species. We noted how the worst discrimination occurred in species with open nests and clustered colonies and we suggested that in these species visual communication of the individual status could also be possible (Turillazzi et al. 2004). Alkenes proved to be the compound most involved in the Discriminant Functions even in those species (*L. vechti* and *E. fraterna*) where branched alkanes are present. We concluded that fertile and non-fertile individuals are characterised by their cuticular chemical profiles in at least two species of hover wasps (*P. striatula* and *E. fraterna*) and that the differences could be due to factors influencing metabolism of these compounds, age or exposition to environmental substances during external activities.

Subsequent investigations confirmed similar findings also in *P. mellyi* where individuals with more developed ovaries present a higher percentage of branched alkanes and a lower percentage of n-alkanes with respect to individuals with less developed ovaries (Turillazzi et al. 2008).

5.3.2.2 Alarm

It is possible, but so far not demonstrated, that individuals of a colony can be informed about predator attack or an alien individual by the alarm behaviour of other individuals. Christina Coster-Longman and Angelo Fortunato performed preliminary but interesting experiments on alarm propagation in a cluster of colonies of *Liostenogaster vechti* built on the ceiling of a gazebo at the border of a forest. They attached a cardboard screen to the ceiling in order to divide the group of nests into two parts, so that the individuals on one side could not see a dead hornet, fixed to the tip of a stick, which the experimenters slid along the ceiling towards the colonies on the other side of the screen. Several times it was observed that the individuals on the un-challenged side started to become agitated and flew from their nests shortly after individuals in the challenged side manifested an alarm reaction to the dead hornet, many of them looping en masse down from their nests and back so that they became visible to the wasps on the other side of the screen.

In some Vespidae, as well as in several species of bees and ants, volatiles found in the venom secretion have been shown to serve a pheromonal function. In particular venom has been reported to function as an alarm pheromone not only in various species of the genus *Vespa* (Saslavasky et al. 1973, for *V. orientalis*; Veith et al. 1984 for *V.crabro*) and *Vespula* (Aldiss 1983 for *V. vulgaris*; Heath and Landolt 1988 for *V. squamosa*; Landolt et al. 1995 for *V. maculifrons*) but also in several species of the genus *Polistes* (Jeanne 1982 for *P. canadensis*; Post et al. 1984 for *P. exclamans* and *P. fuscatus*; Bruschini et al. 2006 in *P. dominula*; Bruschini et al. 2008 in *P. gallicus*) and in the swarming species *Polybia occidentalis* (Jeanne 1981). The volatile fraction of venom from seven species of hover wasps was analysed by Dani et al. (1998) who found a mixture of linear alkanes and alkenes with chain length ranging from C_{11} to C_{17} with clear differences among the various species. In *P. jacobsoni* and *P. mellyi* analyses also showed the presence of some spiroacetals, which in *Polistes* and *Vespa* seem to function as alarm pheromones (reviewed by Francke and Kitching 2001). In any case, behavioural bioassays to demonstrate this function of the venom in the Stenogastrinae have, for the moment, furnished negative results (Dani et al. 1998, personal observations).

5.3.2.3 Sexual Interactions

Great part of this argument is treated in Chap. 3; here I just wish to remind the reader that visual communication is utilised in sexual interactions of the two sexes or in competition between males. The display of some coloured body parts, such as the white stripes of flying males of various species of *Parischnogaster* and *Metischnogaster*, are honest signals used in contests which tend to leave only a few individuals on the spot at the time when mating will take place on the arrival of the females. White stripe display can be considered an honest signal of strength as it implies great energetic cost for stretching the gaster to its full extension. We stressed the fact that the stripe display can also represent a status badge as males "charged" with an extra stripe painted on the second gastral tergite (and hence always visible), are more often challenged by other males and have to restore themselves, feeding on feeders in a flying cage, more often than unpainted males (Beani and Turillazzi 1999).

Chemical communication is also probable as the extension of the gaster uncovers some areas of the cuticle that are full of the exit holes of clusters of tegumental glands which are supposed to secrete substances having a communication function in the mating system of the species. Males of all species of hover wasps studied so far, present glands of this type positioned roughly in the same areas of their gastra [(Turillazzi and Calloni 1983; Turillazzi and Francescato 1989) (see Fig. 2.21)]. Mandibular glands of males could be another source of possible sexual pheromones, in fact in some species (*L. flavolineata*, Delfino et al. 1998) these glands are much larger than those of females (Fig. 2.22). In this regard, it is worth remembering that, while patrolling, the males of *Parischnogaster* often keep their mouth pieces extended (unpublished observation) as if they were releasing some pheromonal substance; but this must be confirmed and tested.

We also observed that towards the end of patrolling males (but also females during other times of the day) will discharge their hindgut contents on leaves or other substrata which, in most cases, are the patrolling landmarks. Considering the number of sexual interactions observed on these sites we hypothesised they represent "hotspots" marked for facilitating species recognition and sexual advertising in the jungle environment (Beani et al. 2002). Actually GC–MS analyses of the spots laid on the landmark leaves revealed the presence of mixtures of long chained hydrocarbons similar to those of the cuticle and of the hindgut contents of the wasps (Beani et al. 2002).

We still lack any information about possible sexual chemical messages released by females with the exception of the above mentioned deposition of hindgut contents. In *Polistes fuscatus*, *P. exclamans* (Post and Jeanne 1983, 1984) and in *Belonogaster petiolata* (Keeping et al. 1986) venom has been shown to have an attractive effect on males, but this has never been confirmed in other species of social wasps nor has it been tested in any species of Stenogastrinae.

References

Aldiss JBJF (1983) Chemical communication in British social wasps (Hymenoptera: Vespidae). Ph.D. Dissertation, University of Southampton, UK

Ayasse M, Marlovits T, Tengö J, Taghizadeh T, Francke W (1995) Are there pheromonal dominance signals in the bumblebee *Bombus hypnorum* L (Hymenoptera, Apidae)? Apidologie 26:163–180

Baracchi D, Petrocelli I, Cusseau G, Pizzocaro L, Teseo S, Turillazzi S (2012) Facial markings in the hover wasps: quality signals and familiar recognition cues in two species of Stenogastrinae. Anim Behav in press

Beani L, Turillazzi S (1999) Stripes display in hover wasps (Vespidae-Stenogastrinae): a socially costly status badge. Anim Behav 57:1233–1239

Beani L, Sledge MF, Maiani S, Boscaro F, Landi M, Fortunato A, Turillazzi S (2002) Behavioural and chemical analysis of scent-marking in the lek system of a hover-wasp (Vespidae, Stenogastrinae). Insect Soc 49:275–281

Bernard J, Brennan BJ (2007) Abdominal wagging in the social paper wasp *Polistes dominulus*: behavior and substrate vibrations. Ethology 113:692–702

Breed MD (1998) Chemical cues in kin recognition: criteria for identification, experimental approaches, and the honey bee as an example. In: Vander Meer RK, Breed MD, Espelie KE, Winston ML (eds) Pheromone communication in social insects: ants, wasps, bees and termites. Westview Press, Colorado

Breed MD, Julian GE (1992) Do simple rules apply in honeybee nestmate discrimination? Nature 357:685–686

Breed MD, Diaz PH, Lucero KD (2004) Olfactory information processing in honeybee, *Apis mellifera*, nestmate recognition. Anim Behav 68:921–928

Brennan BJ (2007) Abdominal wagging in the social paper wasp *Polistes dominulus*: behavior and substrate vibrations. Ethology 113:692–702

Bridge C, Field J (2007) Queuing for dominance: gerontocracy and queue-jumping in the hover wasp *Liostenogaster flavolineata*. Behav Ecol Sociobiol 61:1253–1259

Bruschini C, Cervo R, Turillazzi S (2006) Evidence of alarm pheromones in the venom of Polistes dominulus workers (Hymenoptera: Vespidae). Physiol Entomol 31:286–293

Bruschini C, Cervo R, Turillazzi S (2008) Nesting habits and alarm pheromones in *Polistes gallicus* (Hymenoptera Vespidae). J Insect Behav 21:123–129

Bruschini C, Cervo R, Turillazzi S (2010) Pheromenes in ocial wasps. In: Litwack G (ed) Vitamins and hormones. Academic Press, Burlington, pp 447–492

Cervo R, Turillazzi S (1989) Nest exchange experiments in *Polistes gallicus* (L.) (Hymenoptera, Vespidae). Ethol Ecol Evol 1:185–193

Cervo R, Dani FR, Turillazzi S (1996) Nestmate recognition in three species of stenogastrine wasps (Hymenoptera Vespidae). Behav Ecol Sociobiol 39:311–316

Cervo R, Dani FR, Turillazzi S (2000) Chemically based recognition in a stenogastrine wasp: *Liostenogaster flavolineata* (Hymenoptera, Vespidae). Insect Soc Life 3:19–20

Cervo R, Dani FR, Zanetti P, Massolo A, Turillazzi S (2002) Chemical nestmate recognition in a stenogastrine wasp, *Liostenogaster flavolineata* (Hymenoptera Vespidae). Ethol Ecol Evol 14:351–363

Dani RF, Cervo R, Turillazzi S (1992) Abdomen stroking behaviour and its possible functions in *Polistes dominulus* (Christ) (Hymenoptera Vespidae). Behav Process 28:51–58

Dani RF, Morgan ED, Jones GR, Turillazzi S, Cervo R, Francke W (1998) Species-specific volatile substances in the venom sac of hover wasps. J Chem Ecol 24:1091–1104

Dani FR, Jones GR, Destri S, Spencer SH, Turillazzi S (2001) Deciphering the recognition signature within the cuticular chemical profile of paper wasps. Anim Behav 62:165–171

Dani FR, Jones GR, Corsi S, Beard R, Pradella D, Turillazzi S (2005) Nestmate recognition cues in the honey bee: differential importance of cuticular alkanes and alkenes. Chem Sens 30:1–13

Delfino G, Fortunato A, Formigli F, Turillazzi S (1998) Structural and ultrastructural features of the ectal mandibular gland in Liostenogaster (Vespidae, Stenogastrinae). Insect Soc Life 2:45–52

Destri S, Dapporto D, Fanelli D, Jones GR, Hashim R, Turillazzi S (2002) Preliminary investigations on the role of cuticular hydrocarbons in nestmate recognition of *Parischnogaster mellyi* (Hymenoptera Vespidae). Insect Soc Life 4:145–150

Detrain C, Deneubourg JL, Pasteels JM (1999) Information processing in social insects. Birkhäuser, Basel, Switzerland

Dreller C, Kirschner WH (1993) Hearing in honeybees: localization of the audiotory sense organ. J Comp Physiol A 173:275–279

Espelie KE, Wenzel J H, Chang G (1990) Surface lipids o f the social wasp Polistes metricus Say and its nest and nest pedicel and their relation to nestmate recognition. J chem Ecol 16 : 2229–2241

Field J (2008) The ecology and evolution of helping in hover wasps (Hymenoptera: Stenogastrinae). In: Korb J, Heinze J (eds) Ecology of social evolution. Springer, Heidelberg, Berlin

Francescato E, Massolo A, Landi M, Gerace L, Hashim RB, Turillazzi S (2002a) Colony membership, division of labour and genetic relatedness among females of colonies of *Eustenogaster fraterna* (Hymenoptera, Vespidae, Stenogastrinae). J Insect Behav 15:153–170

Francescato E, Lunghi K, Turillazzi S (2002b) Acoustic communication in *Eustenogaster* (Stenogastrinae, Vespidae). Insect Soc Life 4:151–158

Francke W, Kitching W (2001) Spiroacetal in insects. Curr Org Chem 5:233–251

Gamboa GJ (1996) Kin recognition in social wasps. In: Turillazzi S, West-Eberhard MJ (eds) Natural history and the evolution of paper wasps. Oxford University Press, Oxford

Gamboa GJ (2004) Kin recognition in eusocial wasps. Ann Zool Fenn 41:789–808

Gamboa GJ, Dew HE (1981) Intracolonial communication by body oscillations in the paper wasp, *Polistes metricus* (Hymenoptera: Vespidae). J Kansas Entomol Soc 54:153–155

Hansell MH (1987) Elements of eusociality in colonies of *Eustenogaster calyptodoma* (Sakagami & Yoshikawa) (Stenogastrinae, Vespidae). Anim Behav 35:131–141

Heath RR, Landolt PJ (1988) The isolation, identification, and synthesis of the alarm pheromone of *Vespula squamosa* (Drury) (Hymenoptera: Vespidae) and associated behavior. Experientia 44:82–83

Jeanne RL (1981) Alarm recruitment, attack behaviour, and the role of alarm pheromones in *Polybia occidentalis* (Hymenoptera: Vespidae). Behav Ecol Sociobiol 9:143–148

Jeanne RL (1982) Evidence for an alarm substance in *Polistes canadensis*. Experientia 38:229–230

Keeping MG, Lipschitz D, Crewe RM (1986) Chemical mate recognition and release of male sexual behavior in polybiine wasp *Belonogaster petiolata* (Degeer) (Hymenoptera: Vespidae). J Chem Ecol 12:773–779

Keller L, Nonacs P (1993) The role of queen pheromones in social insects: queen control or queen signal? Anim Behav 45:787–794

Klahn JE, Gamboa GJ (1983) Social wasps: discrimination between kin and non-kin brood. Science 221:482–484

Krebs JR, Davies NB (1987) An introduction to behavioural ecology, 2nd edn. Blackwell, Oxford

Krombein KV (1991) Biosystematic studies of Ceylonese wasps. XIX: natural history notes in several families (Hymenoptera, Eumenidae, Vespidae, Pompilidae, Crabronidae). Smithsonian Contrib Zool 515:1–41

Landolt PJ, Heath RR, Reed HC, Manning K (1995) Pheromonal mediation of alarm in the eastern yellowjacket (Hymenoptera: Vespidae). Fla Entomol 78:101–108

Liebig J, Peeters C, Oldham NJ, Markstadter C, Hölldobler B (2000) Are variations in cuticular hydrocarbons of queens and workers a reliable signal of fertility in the ant *Harpegnathos saltator*? Proc Natl Acad Sci USA 97:4124–4131

Lunghi K (1999) Architettura del nidi e comunicazione acustica nel genere Eustenogaster. MSc Thesis, Università di Firenze, Italy

Meskali M, Provost E, Bonavita-Cougourdan A, Clément JL (1995) Behavioural effects of an experimental change in the chemical signature of the ant *Camponotus vagus*. Insect Soc 44:347–358

Nieh JC, Tautz J (2000) Behavior-locked signal analysis reveals weak 200–300 Hz comb vibrations during the honeybee waggle dance. J Exp Biol 203:1573–1579

Peeters C, Monnin T, Malosse C (1999) Cuticular hydrocarbons correlated with reproductive status in a queenless ant. Proc R Soc Lond B 266:1323–1327

Post DC, Jeanne RL (1983) Venom source of the sex pheromone in the social wasps, *Polistes fuscatus* (Hymenoptera: Vespidae). J Chem Ecol 9:259–266

Post DC, Jeanne RL (1984) Venom as an interspecific sex pheromone and species recognition by a cuticular pheromone in paper wasps (*Polistes* Hymenoptera: Vespidae). Physiol Entomol 9:65–75

Post DC, Downing HA, Jeanne RL (1984) Alarm response to venom by social wasps, *Polistes exclamans* and *P. fuscatus* (Hymenoptera: Vespidae). J Chem Ecol 10:1425–1433

Pratte M, Jeanne RL (1984) Antennal drumming behavior in *Polistes* wasps (Hymenoptera: Vespidae). Z Tierpsychol 66:177–188

Reeve HK (1989) The evolution of conspecific acceptance thresholds. Am Nat 133:407–435

Ruther J, Sieben S, Schricker B (2002) Nestmate recognition in social wasps: manipulation of hydrocarbon profiles induces aggression in the European hornet. Naturwissenschaften 89:111–114

Sakagami SF, Ohgushi R, Roubik DW (1990) Natural history of social wasps and bees in equatorial Sumatra. Hokkaido University Press, Sapporo

Samuel C (1987) Factors affecting colony size in the stenogastrine wasp *Liostenogaster flavolineata*. PhD thesis, University of Malaya, Kuala Lumpur

Saslavasky H, Ishay J, Ikan R (1973) Alarm substances as toxicants of oriental hornet, *Vespa orientalis*. Life Sci 12:135–144

Seeley TD (1989) The honey bee colony as a superorganism. Am Sci 77:546–553

Singer TL, Espelie KE, Gamboa GJ (1998) Nest and nestmate discrimination in independent-founding paper wasps. In: Vander Meer RK, Breed MD, Winston ML, Espelie KE (eds) Pheromone communication in social insects: ants, wasps, bees, and termites. Westview Press, Colorado

Sledge MF, Fortunato A, Turillazzi S, Francescato E, Hashim R, Moneti G, Jones GR (2000) Use of Dufour's gland secretion in nest defence and brood nutrition by hover wasps (Hymenoptera Stenogastrinae). J Insect Phys 46:753–761

Sledge M, Boscaro F, Turillazzi S (2001) Cuticular hydrocarbons and reproductive status in the social wasp *Polistes dominulus*. Behav Ecol Sociobiol 49:401–409

Starks PT (2004) Recognition systems: from components to conservation. Ann Zool Fenn 41:689–690

Tibbetts EA (2002) Visual signals of individual identity in the wasp *Polistes fuscatus*. Proc R Soc Lond B 269:1423–1428

Turillazzi S (1990) Social biology of *Liostenogaster vechti* Turillazzi 1988 (Hymenoptera Stenogastrinae). Trop Zool 3:68–87

Turillazzi S, Calloni C (1983) Tegumental glands in the third gastral tergites of male *P. nigricans* and *P.mellyi*. Insect Soc 30:455–460

Turillazzi S, Francescato E (1989) Observation on the behaviour of male stenogastrine wasps (Hymenoptera, Vespidae, Stenogastrinae). Actes Coll Insectes Soc 5:181–187

Turillazzi S, Sledge MF, Dapporto L, Landi M, Fanelli D, Fondelli L, Zanetti P, Dani FR (2004) Epicuticular lipids and fertility in primitively social wasps (Hymenoptera Stenogastrinae). Physiol Entomol 29:1–8

Turillazzi S, Fanelli D, Theodora P, Lambardi D, Ortolani I, Hashim R, Baracchi D (2008) Determinants of immature brood and nest recognition in a stenogastrine wasp (Hymenoptera Vespidae). Ethol Ecol Evol 20:17–33

Vander Meer RK, Morel L (1998) Nestmate recognition in ants. In: Vander Meer RK, Breed MD, Winston ML, Espelie KE (eds) Pheromone communication in social insects: ants, wasps, bees, and termites. Westview Press, Colorado

Veith HJ, Koeniger N, Maschwitz U (1984) 2-Methyl-3-butene-2-ol, a major component of the alarm heromone of the hornet, *Vespa crabro*. Naturwissenschaften 71:328–329

West-Eberhard MJ (1969) The social biology of polistine wasps. Misc Publ Mus Zool Univ Michigan 140:1–101

Zanetti P, Dani FR, Destri S, Fanelli D, Massolo A, Moneti G, Pieraccini G, Turillazzi S (2001) Nestmate recognition in *Parischnogaster striatula* (Hymenoptera Stenogastrinae), visual and olfactory recognition cues. J Insect Phys 47:1013–1020

Chapter 6
The Nest of Hover Wasps

6.1 General Characteristics of Social Wasp Nests

The nest is a character common to all social insects and represents an important factor in the origin and evolution of their social life. A nest means a space, in part isolated from the external environment, where social life takes place: here the immature brood is reared and protected by the adults as well as by the features of the nest itself. According to Hansell (1996), the nest provides its builder with an element of control over the environment; the creation of stable conditions restrains the foundation of new colonies since it encourages young individuals to remain on the parental colony. In turn, the colony increases in size and interactions between colony members become more complex. At the same time the nest, by providing protection against predators and parasites, facilitates certain types of communication between colony members and improves colony capacities for rearing the immature brood through storage of alimentary reserves. On the other hand, it limits the biology of its builders, due to an increase in their direct dependency on the nest itself.

In social insects the nest can be dug in the soil or inside rotting wood, made of living vegetal tissues, as in the case of galls, or consists of elaborate structures constructed from a wide range of materials. In all cases it is an essential component of colony life and is the result, and sometimes the cause, of some of the characteristics of the colony.

Animal artefacts are considered to be a manifestation of their extended phenotype (*sensu* Dawkins 1982) and a record of part of their behavioural repertoire (frozen behaviour, Emerson 1938). The nests of social wasps, like those of other social insects, are functional for the organisation of their colonies and represent the result of an evolutionary process conditioned by several selective forces.

The nest architecture of social wasps is an extremely interesting field of study, because in these insects nest structure reaches the highest degree of complexity and variability in the whole of the animal kingdom. Moreover, wasp nests exhibit

S. Turillazzi, *The Biology of Hover Wasps*, DOI 10.1007/978-3-642-32680-6_6,

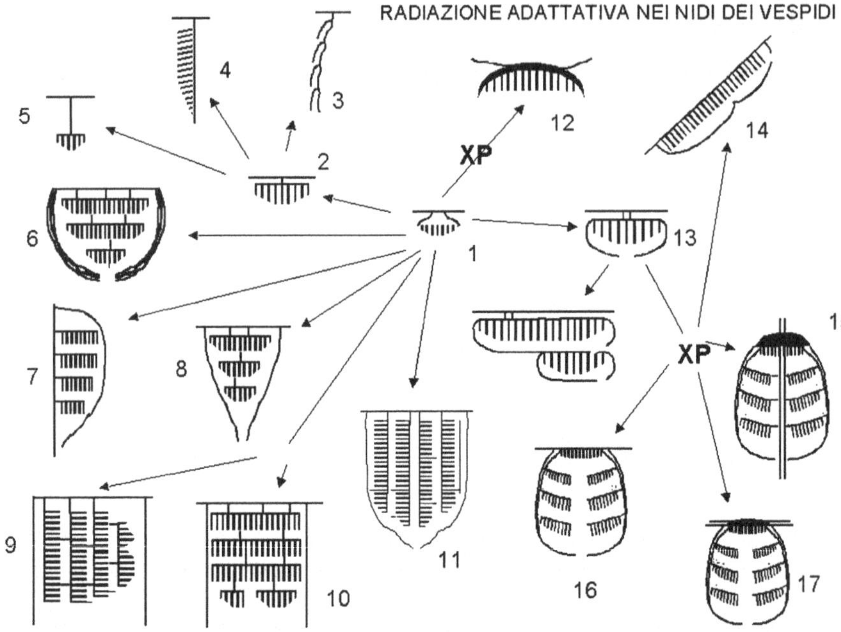

Fig. 6.1 A simplified table developed on the one proposed by Jeanne (1975, p. 179) for the evolution of nest architecture in social wasps. The author mainly takes in consideration Polistinae and Vespinae wasps. From a basic, almost pedunculate, shape (1), more complex designs originated with an increase in size and two main "inventions": that of the peduncle and that of the envelope. No 6, 7, 8 refer to basic architecture of vespine genera while no 2–5 to independent-founding Polistinae and no 9 to 16 to swarm-founding Polistinae (some Ropalidinii and all the Epiponini). XP indicates a secondary loss of the nest peduncle

certain characteristics which make them unique among all social insects, in particular their wide range of shapes and architecture.

The paper by Jeanne (1975) on the evolution of nests in the Vespidae remains a classic in this specific subject (Fig. 6.1 simplifies the main evolutionary steps); the author considers the nest to be the result of various selective pressures.

The immature brood of social wasps and other social insects is extremely vulnerable. Their limited mobility, the large number of larvae packed together and food storage all make the nest a highly attractive food source for many predators. The brood thus necessarily relies on the adults to protect them if they are to survive. This type of defence is direct and accounts for the weapons of the adults (stings, venomous or irritant secretions). But defence can also be indirect when it is afforded by protective structures provided by the nest itself (such as the peduncle and envelope). Therefore the nature of the main selective forces affecting nest architecture can be internal or external to the colony. The most important are:

- Size and morphology of the adults
- Protection of the colony from the elements and the need to create a homeostatic environment

- Defence from predators, parasites and pathogens
- Necessity to optimise rearing the immature brood
- Characteristics and economy of construction material
- Necessity to assure optimum communication between individuals within the context of the colony.

On the other hand, while these factors do indeed influence nest architecture, the nest itself can determine the conditions for the evolution of various colonial features, including the level of social organisation (Hansell 1996).

The shape of a given nest can be considered the result of the building behaviour of several individuals (Dawkins 1982; Wenzel 1996) but in the independent founding species, which include all the primitively eusocial Stenogastrinae and the highly social Vespinae, some of the basic characteristics are determined by the behaviour of a single individual: the foundress female. These features are affected by the choice of the nest foundation site, the point of attachment of the nest and the construction of the embryo nest. In species where a swarm of wasps founds the nest, such choices are determined by the auto-organisation of the behaviour of several individuals.

The nest of social wasps derives from the nests of solitary species: the nest of most Eumeninae—the potter wasps—is made of mud, but some species use vegetable material, collected from decomposing wood, to build their cells. For example *Zethus miniatus* which exhibits a basic form of sociality (West-Eberhard 1978), builds mimetic nests which resemble bundles of vegetable trash.

Social wasps build almost exclusively with vegetable material, with some exceptions, and in this respect differ from the most socially evolved species of bees. The nest is mainly built from material which the wasps do not produce themselves but collect from the environment. The material, with the addition of oral secretion, is then treated in the mouthparts, assembled together and modelled to form standard units, the cells, where only one larva is reared at a time up to the transition to the adult stage. In this respect, all wasps differ from certain bees, such as bumblebees, where some individuals are reared in common cells.

To gather good vegetable material, wasps need good instruments: good, strong paper is made from long wood fibres which are not always easily available. So some wasps collect plant hairs while others engage in cutting or scraping fibres from seasoned wood. Strong and sharp mandibles are necessary for this task. Potter wasps are not so well equipped from this point of view, nor are the hover wasps. The other component necessary for good building material is the cement which binds the material together. In wasps, this is the oral secretions which, emitted by the salivary glands, solidify on contact with the air. The paper is thus applied as a soft paste and modelled by the mouth parts of the insects: the better (more adhesive) the cement, the stronger the paper. In some wasps, like some species of *Ropalidia*, the cement is so important that some structures of the nest (for example the envelope in *Ropalidia opifex*, Maschwitz et al. 1990) are built with this material alone.

If we compare the nests of Polistinae and Vespinae wasps, it can be seen how the latter group maintains a high but quite uniform architectural standard in nest construction, with piles of combs enveloped in a protection which is quite similar across the species. The Polistinae, on the contrary, display a wider range of variability: this occurs especially at the subfamily level (which comprises as many as 26 genera) while variability is limited at the single genus level. For example, although the genus *Polistes* is one of the largest in the subfamily, their nest architecture is very uniform, with nests composed of a single sub-circular comb of cells very similar in all the species. It should be added that in various genera of Polistinae the nest is built by swarms of individuals in a fashion of building which Jeanne and Bouwma (2002) call "explosive nest construction" compared with the "progressive nest construction" of independent founding species. In the first type, the workers build much of the new nest in a very short time, even before the queens arrive; in the second type each new cell is added just before the queen lays an egg inside it.

The study of both intra and inter-specific variability in the architectural responses produced by social wasps is particularly interesting in stenogastrine wasps. These wasps, which are characterised by limited colony size and independent foundation, present an extremely wide and complex range of architectural solutions.

6.2 The Nests of Stenogastrinae

6.2.1 Nest Architecture

Stenogastrine nest architecture differs from that of other social wasps and has such an incredible variety of shapes (Ohgushi et al. 1983) that it has in some instances been used as a systematic character. In fact, especially for a field student or collector, is practically impossible to recognise some species relying on the morphological characters of the wasp alone, yet sometimes it can be quite easy if the shape of the nest is considered as well. This peculiarity in the nest architecture of Stenogastrinae was recognised by the very first entomologists who studied them; Pagden (1958) divided them into four groups according to the shape of their nests and Sakagami and Yoshikawa (1968) distinguished the new species *Eustenogaster calyptodoma* from *E. micans* on the basis of important differences in nest architecture.

None of the nests of any stenogastrine species has a petiole (or peduncle), which is one of the most striking differences with respect to the nests of other eusocial wasps. Rather, cells are built directly onto various kinds of flat or threadlike substrates. In other social wasps, the peduncle is an extremely important structural part of the nest: it permits constancy in the very first part of construction because it "standardises" attachment to the substrate. Perhaps one of the most important reasons for the wide variability of nest architecture in hover wasps is simply the absence of a petiole. Not all other social wasps build their nests with a peduncle, but it is certainly an ancestral character of the Polistinae+Vespinae; it is secondarily absent in the nests of various genera of the tribe Epiponini (e.g. *Apoica*) of the subfamily Polistinae.

Another important difference is that the construction material used by the Stenogastrinae is of "poor quality" compared with that used by other social wasps. According to Hansell (1987b) this is a direct consequence of the fact that their mandibles are not suitable for collecting long vegetable fibres, unlike those of other vespids (see Fig. 6.70 and 6.71). Moreover, the collected material is held together by a salivary matrix that is too limited in quantity, and probably in quality, to assure the production of a good paper paste. Hansell maintains that this limits the possibility of building large nests and, consequently, the formation of large colonies, which in turn prevents the development of a more evolved social life in these wasps (Hansell 1987b).

Even if it is evident that other characteristics of the biology of these wasps have limited their social evolution (see Sect. 7.5.7), it is clear that the construction material is somewhat inferior and even more variable than that of the other social wasps. The sum of these two factors seems to further account for the wide variability in nest architecture which can be observed in the hover wasps.

Another general feature which characterises the nests of these wasps is that the larvae, when ready to pupate, do not spin a complete cocoon to seal their cell openings as the other social wasps do. Actually, in several genera the pupal cells are closed by the adults with the same kind of material used for building the other parts of the nest. In the species of "*Parischnogaster striatula* group" (Ohgushi et al. 1983) the cells are not closed at all, probably because the envelope affords sufficient protection and provides a suitable microhabitat for the development of the brood, consequently saving the collection of more building material needed for the opercula. The same technique of narrowing the alveolar opening is also found in all the species belonging to the genus *Eustenogaster* (Hansell 1987a; Krombein 1991). In this genus the cells are also protected by a nest envelope. The same is true for the known nests of species belonging to the genus *Stenogaster* (Spradbery 1975; Hansell and Turillazzi 1990).

6.2.2 *Intra and Interspecific Variability*

Interspecific variability in nest architecture is certainly high in these wasps but considerable variability also occurs in some single species. Within-species variation in nest architecture is evident, for example, in *Parischnogaster mellyi* where nests vary from comb-like to linear, depending on substrate length (Hansell 1981). In the case of *Parischnogaster alternata* intra-specific variation would undoubtedly allow maximum exploitation of limited nesting sites, given the scarce availability of natural suspensions and their inability to build pedicels (Coster-Longman and Turillazzi 1995).

Here I present genus by genus the characteristics of nests of the species for which we have sufficient information: it should be borne in mind that we do not yet know the architecture of any species belonging to the genus *Cochlischnogaster*. We also only have scarce information on the nest architecture of the two Papuan genera (*Stenogaster* and *Anischnogaster*), while we have more detailed information for the most common species of the Eastern Region genera.

6.2.3 *Stenogaster*

In this genus, distributed in the Papuan Region and consisting of 11 species (revised by Van der Vecht 1975), we know only the mud nest of *Stenogaster concinna* described by Spradbery (1975). We also have a rough description of the nest of *S. fulgipennis* given by Lieftinck and reported by van der Vecht (1975).

The nest of *S. concinna* is bell shaped, consisting of a low number of cells and built with "soil" containing a few small pieces of vegetable matter (Spradbery 1975). A nest of the same species, which Mike Hansell and I found in Papua in 1989, has similar characteristics to the nest described by Spradbery (Fig. 6.2), the most important of which is a narrowing of the diameter of the longest cells towards their distal part.

During the same research mission, Mike and I found another probable *Stenogaster* nest with an envelope (or pseudo-envelope), shaped like an inverted flask and with a prominent neck. The nest (built entirely of mud) was formed by a comb of cells supported by a rootlet; an envelope was formed by the continuation of the walls of the external cells, so creating a nest chamber narrowing to an entrance tube running along the rootlet. This species too showed an evident narrowing of the alveolar opening in the most complete cells (Hansell and Turillazzi 1991). The nest, which is shown in Fig. 6.3, had 18 cells and hung from a rootlet under a high earth trench. That was the only nest of the kind we found in our survey and we took particular care in collecting it. It was photographed, collected and secured in a plastic bag that we deposited on the ground at the base of the trench but, unfortunately, one of us fell down causing a small landslide. A stone hit the bag and the nest was in part destroyed.

A seven cell nest of *S. fulgipennis* is described by Lieftinck (in van der Vecht 1975) as a very fragile, cylindrical structure with rounded top, yellowish in colour with the surface covered by "longitudinal ridges and armed with several lanceolate ribbon-like lateral processes nearly as long as the diameter of the nest". The thin support to which the nest was hanging "bore a whitish ring close to the nest" which the descriptor (or van der Vecht himself) interprets as an ant guard.

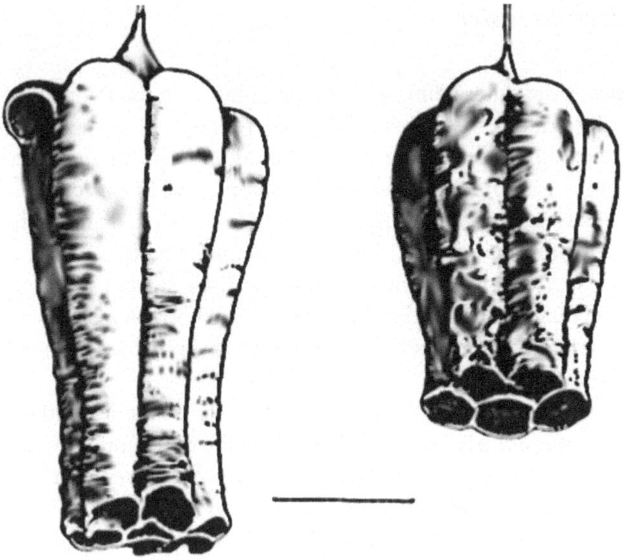

Fig. 6.2 Nests of *Stenogaster concinna* (from Spradbery 1975, re-drawn). Bar = 1 cm

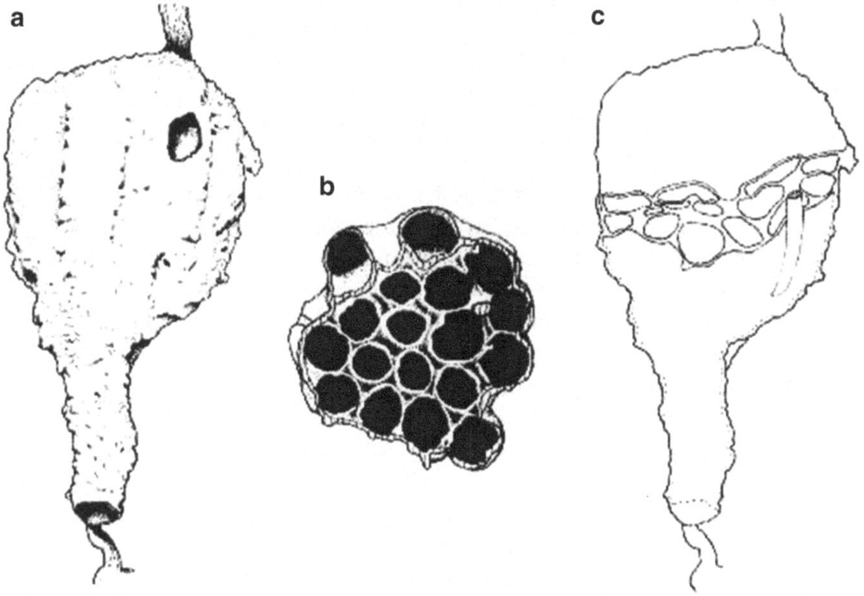

Fig. 6.3 A nest of an unidentified species of *Stenogaster* from New Guinea. (**a**) Drawing of the entire nest. (**b**) Transversal section of the nest at the level of cell openings. (**c**) Schematic view of the nest without the envelope (from Hansell and Turillazzi 1991, re-drawn)

6.2.4 *Anischnogaster*

Anischnogaster is the other genus which is distributed in the Papuan Region. It has been revised by Van der Vecht (1972) who reported five species but gave no information about their nest architecture. Spradbery (1989) furnished a detailed description for the nest of *A. iridipennis* on a sample of 21 nests collected in Papua. The nests of this species are built of fine mud with occasional small pieces of stone and vegetable matter incorporated into the structure. The first cell is attached to a hanging rootlet which is incorporated into the walls. Cells per nest vary between 7 and 18 and, peculiarly, their axis is oriented from horizontal to as much as 45°. When the larvae mature, their cell openings are closed by the adults with the same material they use for the nest.

During the late winter of 1989, my friend Mike Hansell and I studied the biology and social behaviour of hover wasps belonging to the two Papuan genera. We found various colonies of at least four species of *Anischnogaster*, one of which was definitely *A. iridipennis* and the other *A. laticeps*: the nests of the latter were actually quite small with an average of three cells and a maximum of eight. The other two species were probably undescribed (James Carpenter, personal communication) and provisionally termed as *Anischnogaster* sp.A and sp.B. The maximum number of cells found in their nests were 18 and 15, respectively (Fig. 6.4). These small numbers are generally consistent with those of many species of Stenogastrinae but, more to the point, they reflect the particularly small colony size in this genus. The conspicuous feature of the nest material of *A. laticeps* is the rich growth of fungal hyphae ramifying within the cell wall and projecting from both outer and inner surfaces, apparently forming an integral part of the nest material. In appearance, this fungal growth closely resembles the *Fusanum* (Tuberculariaceae) hyphae reported by Krombein (1991) on the nest material of *Eustenogaster eximia* collected in Sri Lanka. Krombein suggests that the fungus helps to bind together the vegetation fragments of the nest material in this species, citing Batra and Batra (1966) who showed that *Termitomyces* fungi help to bind the carton of the nests of subterranean termites such as *Odontotermes* species. In *A. laticeps* it appears that a rich growth of fungal hyphae is a typical feature of the nest, which the wasps apparently make no attempt to control and which larvae and pupae are exposed to and unaffected by. These hyphae supply additional strength to the nest material in the particularly damp, dark sites chosen by this species. It may explain the thinner cell wall of *A. laticeps* nests relative to that of *Anischnogaster* sp.A, which has the same cell dimensions. It therefore appears that this species is trading the loss of nest life expectancy due to fungal decay against shorter term increased strength of cell walls resulting from reinforcement by fungal hyphae.

In *Anischnogaster*, variability between species is also evident in the extent of recycled material from cells from which adults have recently emerged to other cells which contain growing larvae. This is particularly evident in *Anischnogaster* spp. A and B and may result in a mature cell losing up to 50 % of its material after adult

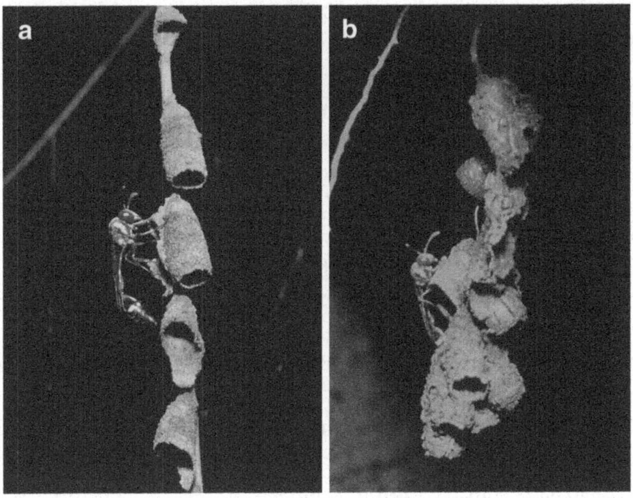

Fig. 6.4 Nests of (**a**) *Anischnogaster iridipennis* and (**b**) *Anischnogaster* sp.B from Papua New Guinea (photos S. Bambi)

emergence. However, material recycling of this kind is virtually absent in *A. laticeps*. We (Hansell and Turillazzi 1995) suggested that the lack of nest material recycling in this species is a mechanism for defending larvae from tachinid parasites (flies which lay their eggs on wasp larvae which will then be killed and eaten by the developing fly larvae). This is also consistent with the observation that in this species new cells reach 60–70 % of their full length before receiving an egg. High levels of tachinid parasitism are also reported in *A. indipennis* and other stenogastrine genera (Spradbery 1989). The absence of nest material recycling in *A. laticeps* therefore appears to be an additional protection associated with its small colony size.

A. laticeps also differs from the other two species in the cap covering the pupal cell. All three species cap the pupal cells with nest material but in *A. laticeps* the operculum is removed after the pupal moult to expose the pupa with its head resting on a ring of dark fungal hyphae, projecting inwards from the walls. In *Anischnogaster* spp. A and B, however, the cap remains intact until broken by adult emergence.

6.2.5 *Liostenogaster*

This genus is considered to be the most primitive of the whole subfamily (Carpenter 1988). This does not seem to relate very well to the social organisation found in some of its species or to the nest architecture which exhibits the greatest diversity of any stenogastrine genus. On the basis of nest design, the whole genus can be divided into two main groups: one builds nests using a prevalence of vegetable material while the other builds nests using mud. Mike Hansell explains this as a secondary adaptation by some species to the predatory pressure of tropical hornets which prey on larvae and pupae of stenogastrine wasps after disrupting the cells built out of paper. In fact I never had the chance to observe a mud nest being preyed upon by *Vespa tropica* but I have repeatedly observed predation on paper nests of various species, not all members of the genus *Liostenogaster*. A student of mine, however, witnessed raids of *Vespa tropica* also on *L. flavolineata* noting that the hornets found more difficulty in crushing the cells than when attacking nests built with vegetable materials (C. Coster-Longman, unpublished observations).

L. flavolineata, *L. pardii*, *L. campanulae*, *L. varipicta*, *L. tutua* and an undescribed species (see below) belong to the first group, which uses mud. The vegetable material nest group, by contrast, contains (at present) *L. nitidipennis*, *L. vechti*, *L. abstrusa*, *L. topographica*, *L. filicis*. The nest of *L. picta* has never been described. Enveloped nests are found in this genus and one species (*L. pardii*) presents the most massive and well-defended nest of all the hover wasps; another species (*L. abstrusa*) builds carton comb nests inside hollow tree branches excavated in part by the wasps themselves (Turillazzi 1999).

One species, *L. tutua*, builds a mud nest of two or more rows of cells placed along a thin twig, which it defends with an ant guard of abdominal substance (see Fig. 6.18) (Turillazzi 1999). In species that use vegetable fibres, we find simple nests with cells scattered or arranged in rows on flat substrates (*L. nitidipennis*: Yoshikawa et al. 1969) (see Fig. 6.19), and more complex ring-like combs with the cell openings facing the centre (*L. vechti*) (Fig. 6.22).

6.2.5.1 *Liostenogaster flavolineata*

Liostenogaster flavolineata, is one of the best-known stenogastrine species (Hansell et al. 1982; Samuel 1987; Strassmann et al. 1994; Field et al. 1998, 1999, 1999) (Fig. 6.5). It uses mud to build its stout and stocky comb nests, which can be found isolated hanging from roots under earth trenches along streams and forest tracks or clustered in dense aggregations attached to the ceilings of suitable nesting places such as bridge vaults, caves and large water pipelines (Fig. 6.6) (Turillazzi 1991). The nesting biology of this species has been thoroughly described by the Malaysian researcher Charlotte Samuel in her PhD thesis which, unfortunately, has never been published. The nests are usually found in places not subject to direct sunlight; measurements of the temperature and humidity at these sites demonstrated the relative constancy of the environment throughout the day (Samuel 1987). The structure of the attachment to the substrate needs an impressive quantity of material if we consider the rest of the nest and the total amount of mud involved in the construction of a medium size nest. The maximum number of cells reported for a nest of this species is 123 (Ohgushi et al. 1990; but authors did not give any further information about this). I counted 110 cells in a specimen I found in Peninsular Malaysia under an earth trench. Another large nest (99 cells) found in the same area weighed over 45 g (Fig. 6.7). The comb is composed of cells with rather thick walls

Fig. 6.5 A female (*left*) and a male of *L. flavolineata* (particulars) (photo by S. Teseo)

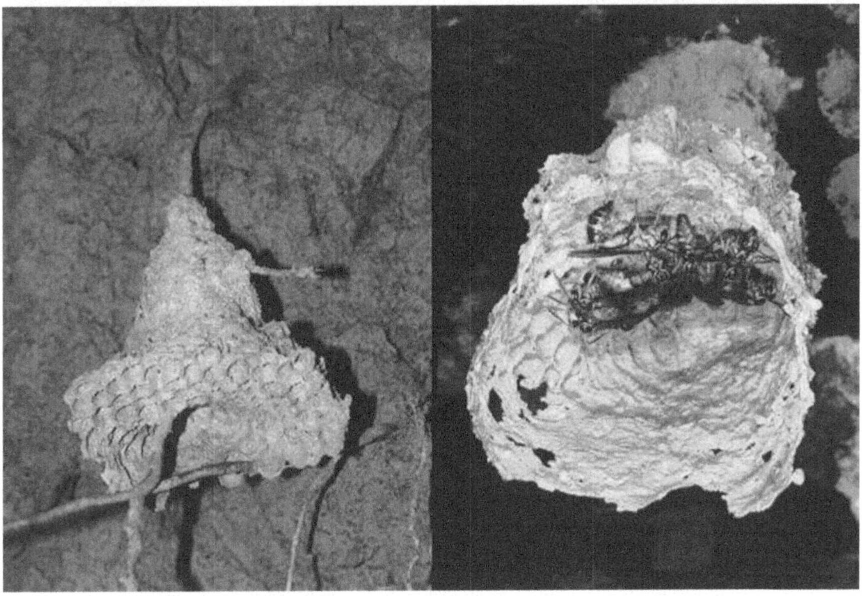

Fig. 6.6 nests of *L. flavolineata* (the *left* one is built on a root and the *right* one is implanted on a flat substrate) (photos by S. Teseo)

(0.5 mm) and an average diameter of 5.5 mm (Samuel 1987); not all the walls are shared completely with peripheral cells and consequently the cells are circular rather than hexagonal units. Usually, some cells in the comb are shorter than others so that the comb surface is often not flat. The insects usually rest in the depressions formed this way, well concealed from an external observer. The concave surface of the comb creates problems in the apposition of contiguous cells due to the occurrence of empty spaces between them (Fig. 6.8).

Fig. 6.7 An earth trench in which a nest of *L. flavolineata* is visible (see the *arrow*) (photo by S. Teseo)

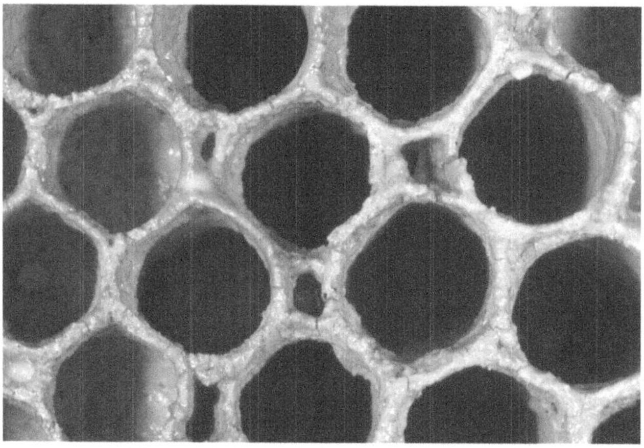

Fig. 6.8 Particular of cells of a nest of *L. flavolineata* (*frontal view*) (photo by S. Teseo)

The external walls of the peripheral cells can be lengthened to form a sort of envelope (defined by Samuel as a "hood") which in some cases is limited to the part of the nest which is not exposed to the light. However, in other cases the whole nest bends towards the light and the envelope is extended fully round the periphery: the result is an almost tubular nest (Fig. 6.9). The hood is usually very thin and breaks

Fig. 6.9 Particular of the "hood" of a nest of *L. flavolineata* (photo by S. Teseo)

easily when the nest is collected. Sometimes the nest has holes, which makes the entire structure look even more like a piece of dry mud hanging from a root (see Fig. 6.60a). In some cases the hoods of adjacent nests can be joined together. Samuel (1987) also observed that in two instances an additional and more internal hood one divided the comb into two parts, each occupied by a different female. The cells can be very long and sometimes host two larvae at different stages of development (Fig. 6.10). The same phenomenon is also reported by the Japanese entomologists Ohgushi, Sakagami and Yamane who found colonies of the same species in Central Sumatra (Ohgushi et al. 1990). These researchers also observed that in *Polistes tepidus malayanus* the same cells are sometimes used for rearing two larvae and in this case the pupae succeed in emerging from the cells without damaging the smaller larvae obstructing their exit (Yamane and Okazawa 1977). Unfortunately in *L. flavolineata* it is not yet clear if both larvae can successfully reach maturity. It seems that cell length reaches a maximum when the nest is implanted on a root and the attachment structure is conical (see Fig. 6.6a). It has also been suggested that the incomplete envelope functions as a landing surface as it opens towards the outside when the nest is built under an earth trench or on the ceiling of a cave (Samuel 1987).

This species can be found isolated but often forms clusters of nests which will be treated in the appropriate section (see later).

Nest material: the material used for the construction of the nest is mud which can be collected from mud-banks in the vicinity of the nest itself (Samuel 1987) or from

Fig. 6.10 A longitudinal section of a nest of *L. flavolineata* showing the extremely long cells and the layer of secretion on the walls of old pupal cells

mud pillars hanging from roots (personal observation). Another source of material can be the nest itself when mud is removed from one part of the nest to build or elongate cells in another, or even from neighbouring nests (Samuel 1987, personal observation). The wasps collect water separately from the mud, from droplets that they find on leaves or twigs—water collection is always performed before mud collection. The colour of the mud is most often yellowish.

Samuel (1987) observed the foundation of nests. The first step involved the smearing of a thin layer of mud on the substrate. Between seven and ten cells may be directly built on a flat substrate while this number falls to 3–5 cells when a thread-like substrate is used. The cells are enlarged as the larvae develop and are closed with a thin layer of mud when they are ready to pupate. In one case I observed nest foundation in captivity where the wasps obtained building material from old nests placed in the rearing cage.

6.2.5.2 *Liostenogaster campanulae*

The nest of this species is formed of a single comb of variable form, usually attached on a vertical flat substrate (Fig. 6.11). When viewed from the side, the comb is triangular in shape with the cell openings facing downward, resembling the shape of a bell (from which the specific name). Nests are implanted on the walls of buildings, caves or rocks or even on tree trunks. The first cells are built with the lateral wall attached to the substrate and their axis parallel to this.

It differs from the nest of *L. flavolineata*, which is almost never implanted on vertical substrata, by having no pseudo-envelope, although both young and mature nests can have various projections of the more external cells which protect the cell

Fig. 6.11 The nest of a mature colony of *L. campanulae* (photo by S. Teseo)

openings of the comb (Fig. 6.12). The cell openings lie on an almost flat plane. Average cell length reaches almost 12.5 mm and the average diameter almost 4.5 mm ($n = 6$) (Bongiovanni 1998).

Nest material. The nest is usually grey-yellowish in colour but dark brown nests have recently been discovered. At Light Microscope level, only rare vegetable particles can be recognised while inorganic particles of various sizes are evident. The material of nests of this species examined by incinerating the nest (Hansell and Turillazzi 1991) showed an average percentage weight of the residue (probably ascribed to inorganic substances) of 92.12 % ($n = 6$; SD = 2.49) (Bongiovanni 1998).

6.2.5.3 *Liostenogaster varipicta*

This species is smaller than the two treated before and builds a nest attached to vertical walls of rocks and buildings (Fig. 6.13). A first row of mud cells with the openings facing downwards is attached directly to the flat substrate while a number of other rows are built directly on the first, forming a comb.

The colour of the nests can vary from grey to yellowish. The alveolar walls are quite thin and shared between adjacent cells, and the openings appear hexagonal. Projections of the more external cells (as found in *L. campanulae*) never occur at any construction stage.

Fig. 6.12 Initial nest of *L. campanulae* with three protrusions which hide the comb

Fig. 6.13 Nest and adults of *L. varipicta* (photo by D. Baracchi)

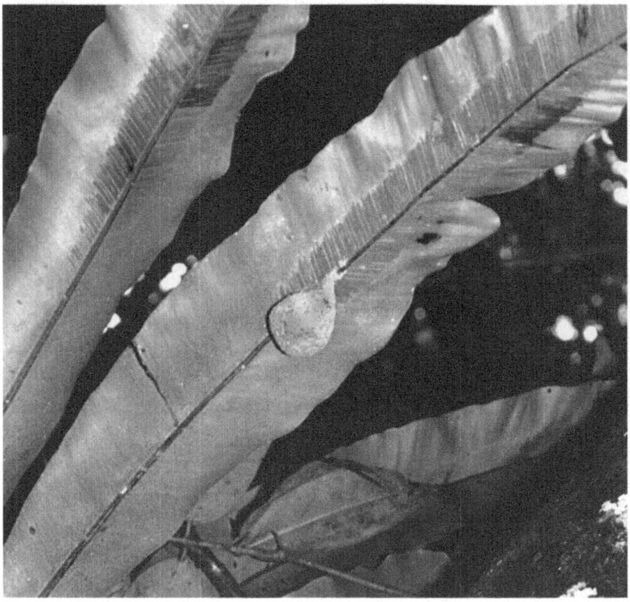

Fig. 6.14 A nest of *L. pardii* under a frond of *Asplenium* fern

6.2.5.4 *Liostenogaster pardii*

Pagden (1958, p. 136) provided a very brief description and illustration of a nest of this species found under a leaf of an *Orchidantha*. I and other entomologists have found nests of this species at several sites in Peninsular Malaysia. The very first one I observed (and that I described as a new species in honour of the great Italian ethologist Leo Pardi who was the first to describe dominance hierarchies in a social wasp) was implanted under the frond of an *Asplenium* fern (Fig. 6.14) at Genting Tea Estate, a private area near Genting Sempah, which lies on the hills near the road from Kuala Lumpur to Bentong, at the border between the states of Selangor and Pahang and at a height of about 650 m. The majority of nests found later were attached to the underside of leaves of trees and plants. However, some nests can be attached to buildings inside or at the edge of the jungle. On plants, the nests are attached to the central vein of the leaves. Nests can be found at heights of 1.5–5 m or more above the ground. Recently we found groups of nests in limited areas; for example six nests of different sizes implanted under the leaves of a fan palm at Hutan Lipur Lentang (160 m asl), a recreation area along the highway from Kuala Lumpur to Karak, and in the same place another group of nine nests was observed in a row of ornamental Ti plants, the dense foliage of which creates a good and protected environment for this species.

The external appearance of a mature nest resembles an almost spherical clay pot. A small circular entrance with an average diameter of 4.49 mm (SD = 0.236, $n = 8$ intact nests) is usually hidden in the upper part of the nest near the juncture

Fig. 6.15 The entrance of a *L. pardii* nest (photo by D. Baracchi)

with the substrate so that it is sometimes quite difficult to realise that the nest is not a solid block of mud (Fig. 6.15). Sections of the collected nests along the plane which includes the base of the nest (i.e. the junction with the substrate) and the entrance hole, reveal the internal architecture (Fig. 6.16); the construction sequence is clear comparing the nests in various phases of development. Nest foundation occurs with the construction of the base of the first cell on a flat surface (Fig. 6.17).

The axis of the first cell is inclined towards the substrate at an angle of about 45°. Lateral cells are certainly added very soon to form a comb. The bases of the cells under the first are displaced with respect to the vertical at the substrate plane and the line from the base of the first cell to that of the lowest one forms an angle of about 116° with the substrate; however, in more mature nests we find that the axis of most cells is almost vertical to the substrate. The envelope begins to be built at the stage of 8–10 cells and derives from prolongations of the external walls of peripheral cells of the comb. The envelope is incomplete in the upper part of the nest, leaving more or less wide access to the cells. Nest enlargement also proceeds with the addition of other cells to the substrate (so enlarging the base of the nest) and is also achieved by adding new cells to the outside of the envelope, as traces of new bases found in young nests indicate. This implies that the envelope must in part be destroyed and then extended to include the new cells. This has been verified on a nest found by Prof. R. Hashim where it is quite evident that the lateral part of the large and fully active nest had been destroyed to enlarge it further. The vestibulum protected by the envelope roughly forms the lower hemisphere in the mature nest. The maximum size reached in a nest with 52 cells measured 48 mm wide × 48.5 mm long × 28 mm tall. Most sizes correlate well with cell number. Weight ranged from a minimum of 0.81 g (a nest with eight cells) to a maximum of 21.73 g (the nest with 52 cells). It is quite astonishing how the shape of the cells in this species is actually hexagonal due to total peripheral wall-sharing and the very thin walls, especially if compared with those of other mud-building species (such as *L. flavolineata* and *L. campanulae*). It is possible that this depends on the presence of the complete envelope which protects the comb.

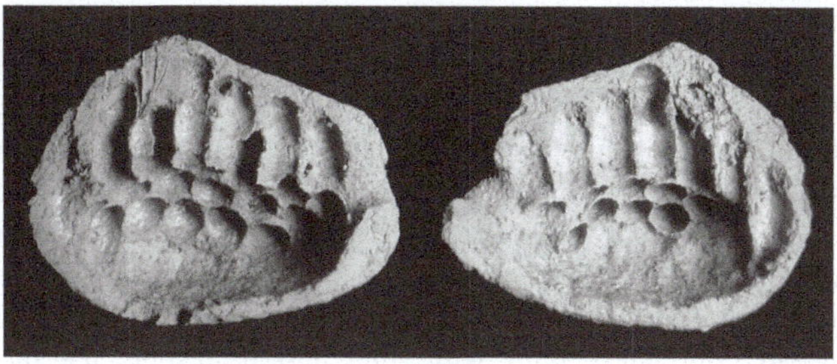

Fig. 6.16 Sectioned nest of *L. pardii*

Fig. 6.17 Construction sequence of the nest of *L. pardii* (from Turillazzi and Carfi 1996)

The reconstructed development sequence of the nest of *L. pardii* (Fig. 6.17) suggests it probably derives from a *L. flavolineata* nest type with an incomplete, thin mud envelope which most likely serves to facilitate landing by returning wasps rather than as a form of defence (Samuel 1987). The defensive function of the *L. pardii* nest seems, in contrast, rather obvious, as a single wasp can guard the nest entrance against ants and other small predators and parasitoids. The thick mud envelope is also very resistant against the ovipositors of parasitoids. But, above all, this nest architecture seems an adaptation to the high predatory pressure from hornets (*Vespa* spp.). This may permit *L. pardii* to nest in open places, more so than the very similar *L. flavolineata*, avoiding the use of mimetic nest structures such as those of other Stenogastrinae. In fact their nests are the most visible of all the hover wasps.

Similar architecture to that of *L. pardii* can be found in some neotropical swarm-founding Polistinae such as *Synoeca*, *Metapolybia*, *Clypearia* (Jeanne 1975), *Occipitalia*, and *Asteloeca* (Wenzel 1991), which build astelocyttarous, calyptodomous nests (that is nests without a pedicel but protected by an envelope) with vegetable material. The nest of *Polybia emaciata* (O'Donnell and Jeanne 2002) is instead completely built with mud and represents the most striking convergence in the Polistinae (also regarding size/weight ratios, as a nest can weigh up to 5 kg) with the nest of *L. pardii*. It is probable that in *P. emaciata* too the use of mud was a defensive adaptation, in this case to serious predatory pressure from raiding ants of the genus *Eciton*.

Nest material. Nests are entirely built of mud. When viewed under the light microscope, mud particles appear extremely small, with scarce granules of quartz never exceeding 0.5 mm. Particles of organic origin are also found, although in very small quantity. All the nests I observed were creamy in colour; they can be easily seen from below on account of their striking contrast with the green of the leaves.

6.2.5.5 *Liostenogaster tutua*

Nests of this species are formed by a row (or more) of cells attached transversely on a small branch, usually from 0.5 to 2 mm in section.

The nests examined ranged from a minimum of 5 to a maximum of 24 cells arranged in 1, 2 or 3 rows. The length of the largest nest reached 71 mm. One of its operculated cells (there were six) measured 9 mm in length with an external diameter of 4 mm. In other nests, some of the cells were arranged in a little cluster. The cells were built out of mud which was yellowish in all the nests examined.

An ant guard of a whitish, jelly-like substance (Fig. 6.18) was found on all the collected nests. The distance of the ant guard from the cells varied between 34 and 20 mm in one uncollected nest. It should be stressed that this seems to be the only species of this genus to use the Dufour's gland secretion to build viscous ant guards to defend the cells.

6.2.5.6 *Liostenogaster nitidipennis*

The photos and drawings published by Iwata (1967) and Yoshikawa et al. (1969) show the nest of this species as an aggregation of mud cells clustered in a multi-layer comb around a rootlet hanging from a cliff or as rows of mud cells attached to the underside of large, green leaves. The species was determined by J. van der Vecht as *Stenogaster nitidipennis* or "a closely related species" (in Iwata 1967) and *nitidipennis* was taken as type species of the new genus *Liostenogaster* van der Vecht in the same paper by Yoshikawa and co-workers. However, the comparison

Fig. 6.18 Nest and adults of *L. tutua* (photo by D. Baracchi)

of some wasps that I collected in peninsular Malaysia with museum specimens assigned to the same species and the provisional key which van der Vecht had prepared for his on-going revision of the genus *Liostenogaster* shed some doubts on the matter. In fact in museum specimens and in the van der Vecht key, *L. nitidipennis* females have a "middle tooth of the mandible truncate or with sharp proximal angle". The wasps I collected, which fit the description of *nitidipennis*, were found on nests of quite different shape and material from the ones reported in the previous papers: they were composed of a set of cells lying on the flat inferior surface of a leaf and sharing their lateral walls to form several rows divided by a space between them, while the material was mainly of vegetable matter. The sharp edge of the second tooth of their mandibles confirms that this species uses scraped material to build their nest. The builders of the mud nests reported by Yoshikawa et al. and by Iwata should have a large, blunt second tooth of the mandible, like all the other species which build mud nests (see Sect. 6.4).

The nest of *L. nitidipennis* is shown in Fig. 6.19. The nests as a result are well camouflaged as the cells are built with dark paper. They are usually built under leaves so that it is quite difficult to find them. The only means I know to find this species is, in fact, to beat the vegetation with a long stick and wait to see if any wasps escape from under the leaves, especially of plants which have large and rigid foliage. In contrast to the similar nest of a species belonging to the *Parischnogaster jacobsoni* group, the cells are arranged more in rows and there is no ant guard.

6.2.5.7 *Liostenogaster vechti*

I named this species (Fig. 6.20) after the famous wasp systematist J. van der Vecht who introduced me for the first time to the systematics of hover wasps and furnished

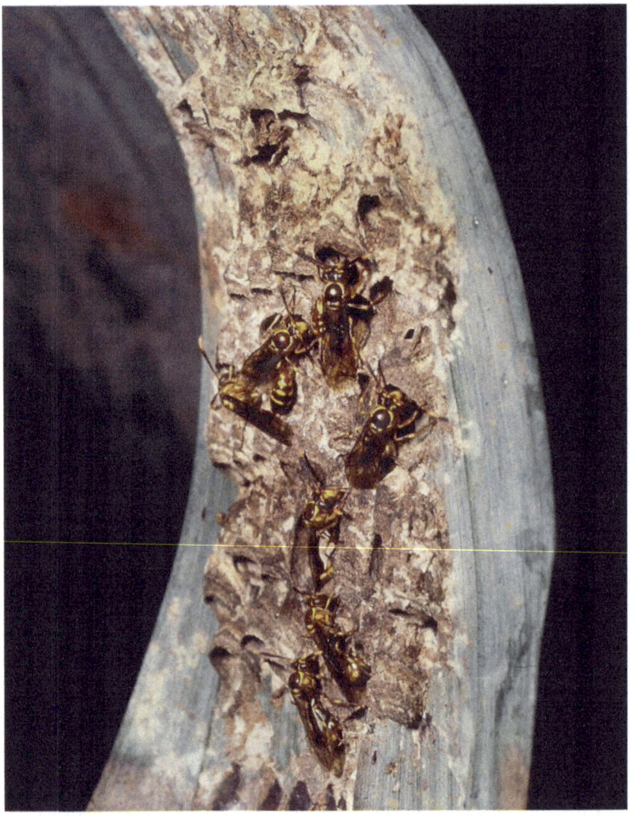

Fig. 6.19 Nest and adults of *L. nitidipennis* on the underside of a leaf upturned for photography

Fig. 6.20 Adults of *L. vechti* on their nest (photo by D. Baracchi) (image is up side down)

Fig. 6.21 Bracket nest of *L. vechti* (photo by D. Baracchi)

me and other researchers with rough notes of his on-going studies on these wasps. In his notes this species was provisionally named as *Liostenogaster arcuata* owing to the form of its nest (Fig. 6.21). *L. vechti* constructs a nest of masticated wood pulp shaped in a single, arched row of up to 40 cells, which, when complete and on horizontal substrata, can form a ring 43–70 mm in diameter (average 56 mm) (Turillazzi 1988a, 1990a). Under crowded conditions, in clusters of colonies, the rings are usually smaller and the wasps can construct a second or even a third story of cells over the first set (Fig. 6.22).

The nests can be built in open spaces such as wooden or metallic poles, tree trunks (Fig. 6.23), exposed rocks, the exterior walls of buildings or in more protected places such as the ceiling of caves, bridges, culverts, etc.

In some areas this species is quite common and conspicuous clusters of nests can be found in great number. Along the 3 km of the road between Bukit Fraser and Jeriau Waterfall, hundreds of nests can be seen. The greatest concentration of nests I have ever found was on the wall of a small water station along the stream which forms the waterfall at an altitude of about 1,100 m. Over 600 nests were present in a single cluster in an area of about 5 $\times$ 3 m. These wasps are quite faithful to their

Fig. 6.22 A ring, double stored, nest of *L. vechti* built on the ceiling of a gazebo (image is up side down)

nesting place and reconstruct their nests quickly if they are destroyed. We have observed a remarkable degree of their persistence on the numerous gazebos built in the recreation area of Jeriau Waterfall. Almost every other year the gazebos are refurbished and re-painted and this operation eliminates the several *L. vechti* clusters of colonies which flourish on the gazebo structures. However, the wasps quickly rebuild their nests and seem to have no problems in using substrates of different colours which change from year to year. Figure 6.24 shows one of these clusters of nests built on a flat, white-painted roof support: the different colour of the nests is evident, indicating that the wasps adopt different material. Nest material is formed of tiny pieces of plant tissues; Ohgushi et al. (1983) distinguish between the reddish brown colour of the cells proper and the whitish colour of the elaborated surface (Fig. 6.21). They also observed that the surface is covered with a pulverised matter. This can vary in colour from dark chocolate to sugary white, to grey and pale green, sometimes forming horizontal "stripes" (Turillazzi 1988a; Coster-Longman 1998). The arena enclosed by surrounding cells, which Coster-Longman calls a "patch", differs from the material used to construct the cells. It seems that females also like other types of material: a ball of adhesive gum experimentally placed in a cluster attracted considerable attention and competition by nearby nest holders which scraped pieces of material from it, eventually adding it to their constructions (Turillazzi unpublished). Some of the nests have thin vertical projections (horn-like processes) starting from the walls of more lateral cells (Fig. 6.25) (Ohgushi et al. 1983; Turillazzi 1988a). These prolongations are usually found only on young nests and it seems they serve as landing pads for the wasps. The wasps garnish the area inside the ring of cells with the application of small stretches of material (see Fig. 6.26) the function of which is unknown. I can only hypothesise that they serve to increase the camouflage of the nests which are almost invisible when constructed on more natural substrates. C. Coster-Longman suggests that they mimic lichens (unpublished observation).

Fig. 6.23 Nests of *L. vechti* attached to a tree trunk

6.2.5.8 *Liostenogaster topographica*

The nest of this species is quite characteristic and always constructed on a flat substrate. It is formed by a central cluster of cells from which lines of material radiate (Fig. 1.13, Fig. 6.27). This confers a reticulated appearance to the surrounding area that looks like a geographical road map. The substrate is always or almost horizontal, and sometimes the nest area can expand to a vertical substrate. In this case the cells are mainly built at the junction between the horizontal and vertical substrate. The substrate is utilised as an integral part of the cell wall. I have counted as many as 83 cells, with a maximum adult population (males and females) of 32 individuals (Fig. 6.28). The maximum diameter of the biggest nest (including the rays) was about 40 cm, but I found an abandoned nest of 60 cm in length and

Fig. 6.24 A cluster of *L. vechti* nests on a recently re-painted gazebo

Fig. 6.25 Particular of the projections from lateral walls in a nest of *L. vechti* (photo by D. Baracchi)

40 cm wide. This species does not seem to be particularly rare and the fact that I have found only a few nests built on substrata other than artificial ones may be evidence of the perfect camouflage these constructions assume in the natural environment. In some other cases we found nests of this species on the underside of plant leaves but probably the most used natural substrates are tree trunks or rocks. Very popular sites used by these wasps for their nests are, again, the gazebos which are so common in

Fig. 6.26 Deposits of material on the pavement of a nest of *L. vechti*

Fig. 6.27 A mature nest of *L. topographica*

resting areas all over Peninsular Malaysia. In this respect they are similar to *L. vechti* with which they can share nesting sites, even if the latter species prefers higher altitudes (usually over 500 m asl up to 1,000 m). However they have also been found on other types of buildings and under the culverts of bridges.

The colour of the nest material is usually white-grey with reddish brown ribs. The ribs (Fig. 6.29) are in fact covered with small reddish brown particles which can also be found scattered over the substrate. Under the microscope, the material seems to be composed mainly of vegetable particles, with additional gross elements. Inorganic particles are also numerous (Baracchi et al. 2009).

Fig. 6.28 Females of *L. topographica* on a mature nest (photo by D. Baracchi)

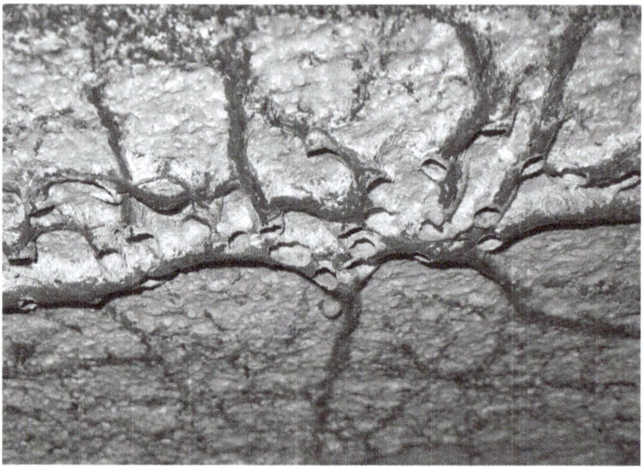

Fig. 6.29 Ribs and cells in a nest of *L. topographica* with different coloured material (photo by D. Baracchi)

6.2.5.9 *Liostenogaster filicis*

The nest of this species may be composed of separate combs (Fig. 6.30) (17, 10 and 7 cells respectively in the nest on which I based the original description of this species, implanted about 10 mm apart on a fern stem 3 mm in section). Some of the cells were attached with their bases to the stem, others were directly attached to the previous ones sharing their walls. One operculated cell was found measuring 11 mm in length and 5 mm in external diameter. The lateral walls of the combs were garnished with pieces of material which contribute to nest camouflage. No ant guard was found even on a second nest of the same species (Fig. 6.31).

Fig. 6.30 A colony of *Liostenogaster filicis*

Fig. 6.31 Cells of a nest of *L. filicis*. The patches of different colour on cell opercula indicate that also the adults of this species practise the extraction of larval peritrophic sacs

Nest material. The nest material of this species is dark reddish brown. Under the Light Microscope it appears to be composed of vegetable matter with particles of various size, vegetable fibres and fungi. Inorganic particles can also be observed. The general shape is similar to the nests reported in pictures by Yoshikawa et al. 1969 and attributed to *L. nitidipennis* (but see what I wrote about this species at Sect. 6.2.5.6); however the latter is clearly made of mud while that of *filicis* is built with paper (Fig. 6.31).

Fig. 6.32 A wasp hovering near the nest of *L. abstrusa* inside a hollow dead branch (photo by D. Baracchi)

6.2.5.10 *Liostenogaster abstrusa*

This species seems quite rare but the particular architecture of its nests certainly contributes a lot to the difficulty in finding it. I have only found four colonies of this wasp, and then completely by chance. In fact I was lucky when I saw a wasp flying out of a dead hollow branch hanging from a tree in the forest at Genting Tea Estate (Fig. 6.32). Nest A was formed of two combs implanted 13 mm apart on the internal wall of this dry hollow branch (about 30 mm in external diameter) belonging to an unidentified plant. The upper comb had eight cells and the lower one has five (Fig. 6.33a). Combs were built on the very first cell which had only half of its surface in contact with the substrate, while the other half bent slightly away. In the lower comb it was particularly evident that the substrate formed part of the wall of the first cell. The combs were inclined at about 40° to the substrate. I was still unsure whether this species built only in very small cavities used as an envelope when I found a second nest. Nest B was composed of only one comb with 11 cells implanted on the internal surface of a hollow dry branch 24 mm in diameter. The longest cells, probably complete, were hexagonal–circular, measuring 13 mm in length and 4 mm in external diameter. In this nest too, only the initial part of first cell was in direct contact with the substrate. Subsequent cells were added both above and below the first cell. There was also a compact prolongation above the apical part of the first cell which formed the base for the lower cell. In the chapter in which I described the species for the first time, I reported that this type of attachment is the only one which is similar to a pedicel (Fig. 6.33b) out of all the described nests belonging to the Stenogastrinae. Actually we can imagine a possible evolution towards a structure similar to a pedicel commencing with a single cell attachment to a flat substrate. It may be that the particular disposition of the nest

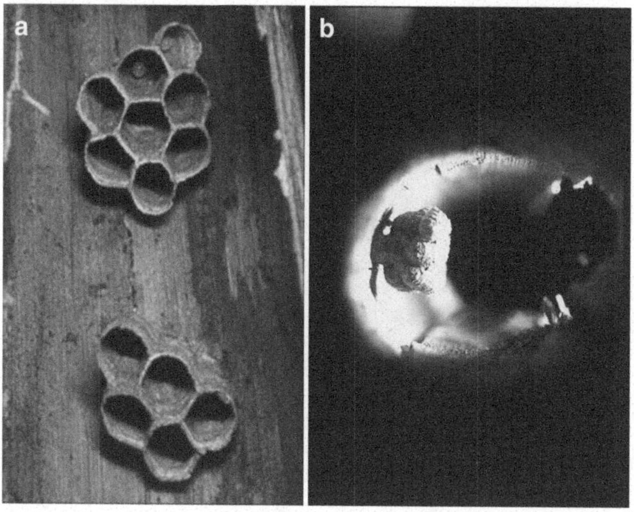

Fig. 6.33 Cells attached to the substrate in a nest of *L. abstrusa*. A: facing the cells. B: side view showing pedicel-type structure

inside hollow branches (which seems a general characteristic of the architecture of this species) favoured attachment based on a single cell. This kind of architecture was confirmed in a further two nests of the same species found in the Gombak area in 2010.

Nest material. Nest material is light brown and whitish in colour; observed under a Light Microscope it consisted almost entirely of vegetable particles with cellular walls and plant hairs. No inorganic particles were evident in this analysis.

6.2.6 Parischnogaster

This genus includes the most common species of hover wasps which have adapted to nesting in rather open habitats, including city gardens and recreation areas. Moreover, some species such as *P. mellyi* or species belonging to the *P. jacobsoni* group seem quite resistant to polluted areas: I have found colonies of the first species hanging from the trees along the highly polluted traffic avenues of the Indonesian city of Bogor; in Malaysia *P. mellyi* seems to be the only species (together with *P. striatula*) still present in the forest park of the Universiti Malaya in the city area of Kuala Lumpur. It probably occurs in other city parks and green areas. Their nests can be found on plants but they are also quite commonly associated with human buildings and it is not unusual to find their colonies hanging from the ceilings of houses (Fig. 6.34) (see also Ohgushi et al. 1988).

Fig. 6.34 Nest of *Parischnogaster mellyi* hanging from a house ceiling (photo by D. Baracchi)

A revision of the entire genus is still lacking, so it is possible that some common species, especially those belonging to the *jacobsoni* group, are as yet unnamed. All *Parischnogaster* females are a shiny black or dark brown colour, while males present white stripes on their gastral terga which are particularly visible during aerial patrolling.

Ohgushi et al. (1983), working on the nest architecture of species belonging to this genus, divided the numerous types of structures into three basic groups: the *mellyi* group, characterised by nests consisting of several irregular combs, either continuous or composed of irregular tiers; the *jacobsoni* group, with nests composed of cells placed in a single or double row on thin supports or scattered over limited flat surfaces such as leaves and always defended by viscous ant guards; the *striatula* group, formed by species with nests hanging from thread-like substrata or implanted on flat surfaces and distinguished from the other groups by the presence of a communal entrance tube surrounded by the cells themselves and by the prolongations of the walls of some cells to form a sort of "pseudo-envelope".

In various papers (Ohgushi et al. 1983; Ohgushi and Salmah 1986; Ohgushi et al. 1988; Ohgushi et al. 1990) the Japanese entomologists provide quite a large collection of drawings of nest types. Unfortunately, as the nests have yet to be attributed to one particular species, more detailed insight into the nesting biology of the genus is for the moment limited, as it is not possible to stipulate nest variation within or between species

Here I report on the nests of the most common species described.

6.2.6.1 *Parischnogaster mellyi*

This species is quite common in wide areas of the Oriental region. Mike Hansell, the British ethologist famous for his books on animal architecture, chose this species for one of the first ethological studies on Stenogastrinae social behaviour (Hansell 1982). This was also the very first stenogastrine I ever came across when I began my studies during a mission to Indonesia in 1979. This species has, probably, the most inconspicuous and flexible (lacking any typical form) nest architecture of all in the subfamily—probably another reason for its success.

The architecture of the nest of *P. mellyi* seems primitive, with cells arranged in irregular combs placed on threadlike substrates such as thin roots (Hansell 1981; Ohgushi et al. 1990, Fig. 7) (Fig. 6.35).

Combs can be initiated separately or contiguously above or below the first comb, thus forming a nest of up to four tiers. The cells are irregular in shape and roundish, sometimes with a limited degree of cell wall sharing. The number of cells in a complete nest can exceed 50 (Fanelli et al. 2005) but Ohgushi et al. (1990) report in a table an extraordinary number of 100 cells omitting to give references.

Hansell (1981) found that in Thailand this species tends to build linear series of cells along hanging substrates. Observations on a population of colonies reared in captivity indicate that in environments where predatory pressure is absent wasps tend to build combed nests with considerable saving of construction material (Coster-Longman 1998) (Fig. 6.36).

The materials this species uses to build its nests can be of different origins but more often it consists of a mixture of organic and inorganic particles. In some cases, however, nests built entirely of mud have been found (Coster-Longman 1998). Nest colour can range from yellowish to dark brown.

This species does not build ant guards for the nest.

6.2.6.2 *Parischnogaster gracilipes*

More regular combs are exhibited by *P. gracilipes* which had been previously included in the sunken genus *Holischnogaster*. This species, which lives in mountain forests where predation from hornets is almost absent, builds unusually large nests for the subfamily, containing up to 54 cells (Hansell 1986) (Fig. 6.37). All nests found by Hansell and myself were attached to fine rootlets underneath eroded

Fig. 6.35 Nest and adults of *Parischnogaster mellyi* (photo by R. Innocenti)

earthen banks or under fallen tree trunks along forest trails in the Mount Kinabalu National Park in Sabah, Northern Borneo. All nests examined were of similar design with the single comb attached by one marginal cell to the suspension and the entire structure was rather elongate and oblique. Cell openings usually face the bank wall. The cells were arranged in rounded combs and the nests I collected were built with a reddish material, including vegetable and mineral particles, while all the nests Hansell found were made with material of plant origin (Hansell 1986).

6.2.6.3 *Parischnogaster nigricans serrei*

Eighty-three nests of this species collected in West Java were studied by my group and the results presented in an unpublished MS thesis by Vellone (1986) who recorded various parameters of their architecture, including certain mechanical characteristics of the material. The nest is of very simple architecture: more or less conical cells arranged in a linear series along a thread-like substratum. The

Fig. 6.36 Nest of *P. mellyi* characterised by various combs along a thread like substrate (photo by D. Baracchi)

material is mainly of plant origin and used cells are often destroyed and their material re-used for building or enlarging active cells. The nest is usually founded by a single female which then begins to apply the ant guard secretion (see later) some cm above the first cell. The second cell is only started after an egg has been laid in the first one.

The substratum used for the nest can vary, natural or artificial, rigid or flexible. Nests have been found implanted on plant stems, lichens, the veins of dead leaves, roots and even on iron threads, cords, electric wires, etc. In case of branching substrata, the nest can follow the shape of the substratum itself. A round cross-section of the substratum is on average 1.25 mm ($\pm$0.19) in diameter($N = 46$). Mature nests consist of up to 38 cells and the average distance between the bottoms of two consecutive cells is 8.22 mm ($\pm$0.15). Fully grown cells are superimposed on the following cells in the line, but only in part. It is quite common to find nests with complete cells paired around the substratum but more unusual to find cells attached to previously built ones but not in contact with the substratum. When all the available substratum has been occupied by a linear series of cells, the wasps

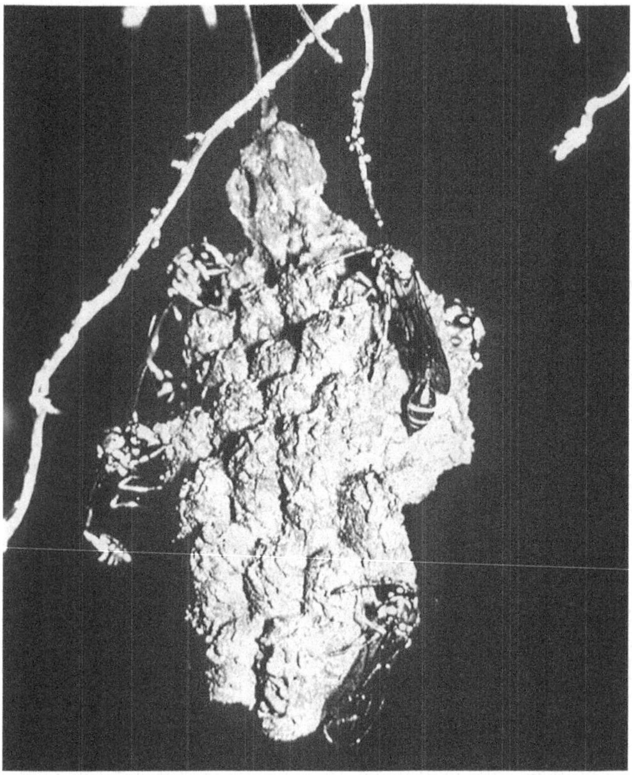

Fig. 6.37 Back of a nest of *P. gracilipes* showing the regular arrangement of the cells in the comb (photo by R. Innocenti)

begin to build more cells opposed to the previous ones so that the number of lines depends on nest age and substratum length. Cells are built slightly rotated with respect to the previous one, so in one particular nest we measured an angle of up 157° between the first and the last cell.

P. nigricans serrei uses only material of plant origin for the construction of its nests.

6.2.6.4 *Parischnogaster jacobsoni*

Other species belonging to the *P. jacobsoni* group (Fig. 6.38) tend to build linear nests with the cells arranged along hanging substrates (or on leaves) and defended by sticky ant guards (Yoshikawa et al. 1969; Turillazzi and Pardi 1982; Ohgushi et al. 1983; Turillazzi 1988b, unpublished). Ohgushi and Salmah (1986, p. 563) illustrate various types of nests apparently belonging to species of this group.

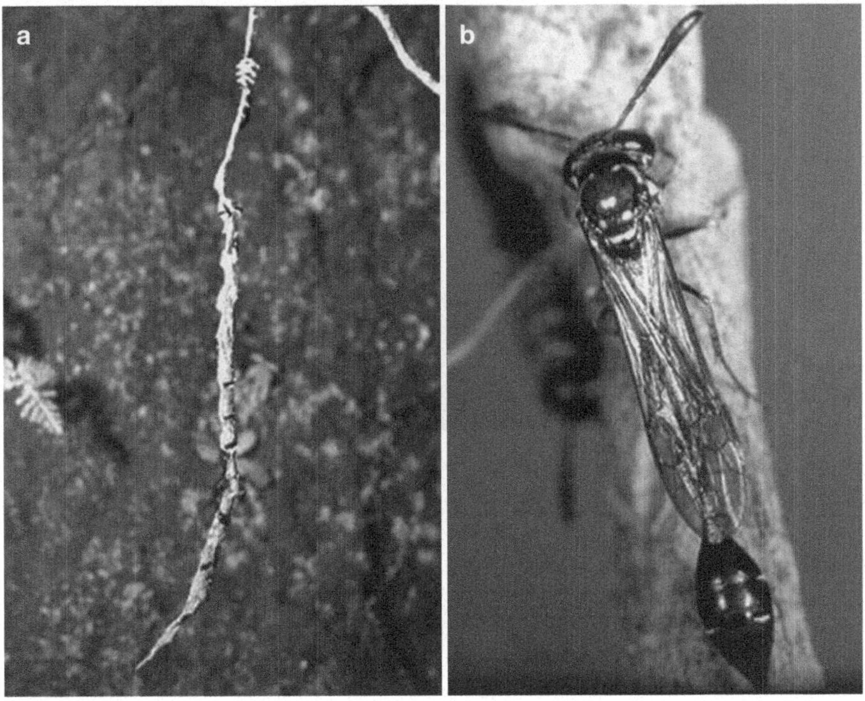

Fig. 6.38 (**a**) Linear nest of a species belonging to the *P. jacobsoni* group. (**b**) A female of *P. nigricans serrei* on a similar type of nest (photos by R. Innocenti)

Nests of the various species are hardly distinguishable from one another and we must wait for the revision of the genus to study possible differences in their architectural characteristics. In general, a linear nest of this type is almost invisible and sometimes camouflaged with leaves or twigs. However some species of the group can implant their cells directly on the flat substrate provided by the lower page of leaves (see Fig. 6.39). In this case the viscous ant guards are placed on the leaf stem. In other cases we observed a clustering of cells especially if the thread-like substrate breaks and available space for nest expansion is limited. These nests correspond to the type Pj2 of Ohgushi et al. 1983).

Construction material, in all cases, seems to consist of plant material.

The nest of one species I determined as *P. jacobsoni* and which I studied in peninsular Malaysia is very similar to that of *P. nigricans serrei* studied in Indonesia but differs from the latter by usually having the alveolar walls extending along the sides of the next cell so that, from side view, the row of cells looks like a saw with smaller teeth than the nest of *P. nigricans serrei*. In *P. jacobsoni* a large percentage of pupal cells are simply narrowed at the alveolar opening, while in the other species all the pupal cells are sealed. The largest nest found in this species had more than 45 used cells for a total length of 48 cm.

Fig. 6.39 Nest of a species belonging to the *P. jacobsoni* group in which the cells are scattered along the inferior face of a leaf (photo by D. Baracchi)

6.2.6.5 *Parischnogaster timida*

The nest of *P. timida* described by Williams (1928) (Fig. 6.40) is extremely similar to the nests of the two described species of *Metischnogaster* (Fig. 6.56). The cells are attached one under the other, the cone-shaped structures above, which deflect raindrops, are built from the same vegetable matter as the nest (Pagden 1962). This represents an interesting case of convergence in construction behaviour unless the author, who described the nest some years later after finding it, erroneously attributed it to the wrong species.

6.2.6.6 *Parischnogaster striatula*

The nests of this species are implanted on thread-like substrata and exhibit a particular rope-like architecture, with cells twisting and opening in a central tube (Williams 1928). Usual nesting sites are sheltered niches along forest roads or trails or under the vaults of caves and man-made buildings where thread-like suspensions are present. Nests exhibit considerable variability in size and external surface. In peninsular Malaysia, my group found various types of nests with the basic structural spiral design. From allozyme analysis performed so far, it seems probable that at least two different forms of nest architecture assigned to *P. striatula* (rough and smooth, Fig. 6.41) are due to within-species variation (Coster-Longman et al. 1996). Additional points favouring within-species variation are that cell length and width are similar in the different types of nests, suggesting the individuals are of similar size. Moreover, the different types of nests contain similar proportions of inorganic residue.

Fig. 6.40 Nest of *P. timida* (redrawn from Williams 1928)

6.2.6.7 *Parischnogaster depressigaster*

The nest of *P. depressigaster* (Rohwer) from the Philippines can be considered as the very first nest of hover wasps whose architecture was described in detail. I wish to report here the description given by Francis X. Williams in his studies on the Philippine wasps (1919) as it renders the characteristics of the structure rather well: "*A very peculiar, fragile and exceedingly slender nest suspended under a bank by one or two stiff black horsehair-like threads—a species of fungus*" and then "*the serpentine passage extended the length of the nest and gave forth vertical pockets or cells used as brood chambers by the wasps . . . and although there is more or less irregularity in cell disposition, the vertical entrances to the cells are almost invariably, if not always, at approximately right angles to the one above and below, and thus alternate openings . . . are in the same vertical plane*" (Williams

Fig. 6.41 Nests of *P. striatula*. *Left*: rough nest. *Right*: smooth nest

1919, p. 168). Williams found several nests, the biggest with 20 cells, hanging under palms, leaves, fallen trees etc. and suggested they were made both with earth and decayed wood. It is probable that this species also lives in Malaysia where we found some very similar nests to the ones described by Williams.

6.2.6.8 *Parischnogaster alternata*

Parischnogaster alternata was distinguished from the very similar *P. striatula* by Sakagarni (in Yoshikawa et al. 1969) mainly by differences in the shape of the nest. This species builds a cylindrical, mixed carton-mud nest (up to 35 cells) reaching approximately 1–2 cm in diameter and 6.5 cm in length (Ohgushi et al. 1983; Turillazzi 1986; Coster-Longman and Turillazzi 1995). It prefers to nest on the ceiling of grottoes, caves and artificial constructions, building its nest on naturally occurring thread-like suspensions (e.g. rootlets) (pointed attachment nests) or implanting them directly onto the substrate which, in this case, is always horizontal (flat-topped attachment nests) (Fig. 6.42).

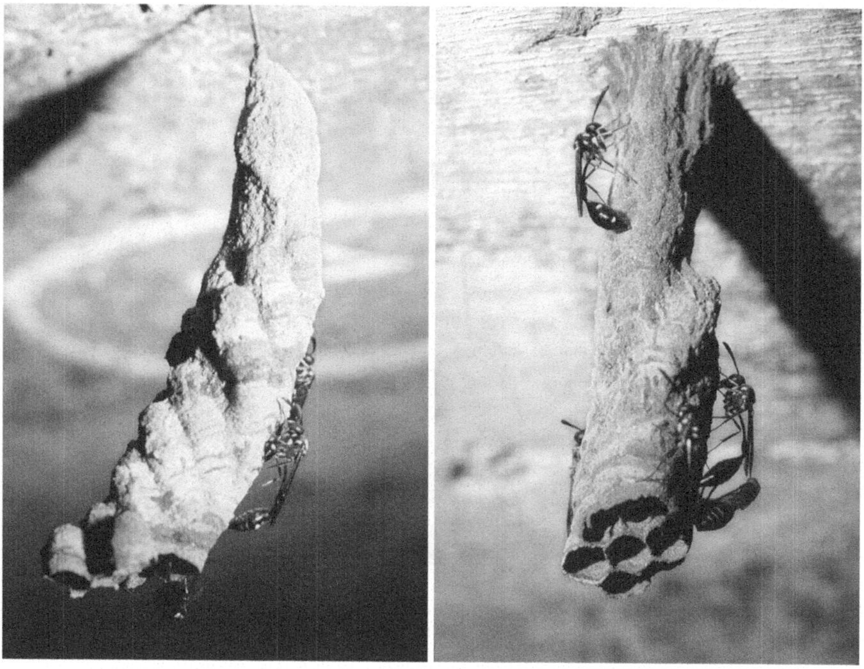

Fig. 6.42 Nests of *P. alternata*. *Left*: "pointed" nest (implanted on a thread like substrate). *Right*: nest directly implanted on a flat substrate (photos by R. Innocenti)

The nest of *P. alternata* may be considered between stage III and IV of the model proposed by Jeanne for the economy of building material and seems more evolved than the primitive astelocyttarous, gymnodomous types (that is nests without pedicel and envelope). It consists of cells which are attached to the previous ones at a slightly lower level in rows. All the cells except the last open downwards into a corridor tube which runs along the side of the nest.

The percentages of inorganic and organic material differ significantly in the two nest types, flat-topped nests containing a higher proportion of mud than pointed nests. This difference in the percentage of inorganic/organic material could depend simply on architectural design according to the type of substrate, but nevertheless is evidence of a particular developed behavioural response in material collection caused by contingency or by a distinct polymorphism in nest building in the same species. This is not the only difference between the two nest types: there is also a significant difference in cell length and width and in the angle the individual cells subtend with the external nest wall Coster-Longman and Turillazzi 1995.

Nests, the structure of which resemble those of *alternata*, have been reported and described by Ohgushi et al. (1983) in a survey on hover wasps performed in Sumatra. The Japanese authors distinguished three more types of these nests (which they designate Ps4, Ps5 and Ps6) two of which are shown in Fig. 6.43.

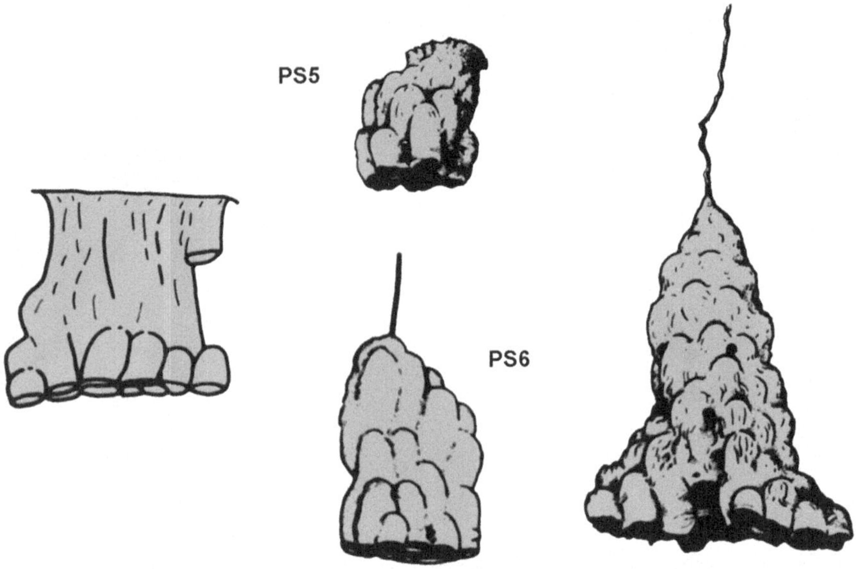

Fig. 6.43 Nest types similar to the basic structure of the nest of *P. alternata* reported by Ohgushi et al. (1983) (redrawn)

6.2.7 *Eustenogaster*

Eustenogaster has recently been revised (Saito and Kojima 2007) and comprises 15 described species, although many others probably still remain undescribed. The genus occurs from Sri Lanka and southern India to Hainan (South China) and Sulawesi (see Fig. 1.3). These are relatively large, robust hover wasps with a distinctive shared nest structure. In this, the central part of the nest is a comb of several (usually 8–20) cells, the number of which appears to be fixed very early in nest development. In all but one known species the peripheral cell walls are extended downwards past the cell mouths to form a narrow spout at least as long as the cells (Fig. 6.44). The space enclosed by these walls (called a "pseudoenvelope") is a sort of "*vestibulum*" which then opens in an entrance tube. In the only known exception, *E. calyptodoma*, a true envelope is formed around the comb, likewise extended down into a spout (Sakagami and Yoshikawa 1968). Variation in nest structure within the genus consists largely of the elaborations of the outer surface of the nest. Nests, while hardly visible, tend to resemble such things as dead leaves, dried fruits or lumps of earth (Krombein 1991; Francescato et al. 2002a), which presumably provides protection against visually hunting predators. Cells are not closed at pupation but only narrowed at their openings; maybe this is a consequence of the protection they have from the *vestibulum* walls.

Ohgushi et al. (1983) divided the nests of this genus into four types: E1, with elongate entrance tube with lacework and large processes on the top; E2, similar to

Fig. 6.44 A nest of *Eustenogaster* sp. showing the characteristic flask-shape architecture (photo by D. Baracchi)

E1 but without ear-like processes and slightly different entrance; E3, similar to a mantis ootheca with simple and short entrance tube and E4, with the envelope completely separated from the comb.

The social behaviour of these wasps is difficult to study owing to the fact that animals spend much of their time inside the *vestibulum*. Hansell destroyed part of the *vestibulum* walls to study the behaviour of *E. calyptodoma* but this was complicated by the fact that the wasps then spent much of their time in repairing the damages (Hansell 1987a). Instead we studied *E. fraterna* by means of small video cameras applied directly to a hole pierced in the *vestibulum* wall. The hole had previously been closed with a black infrared filter and the *vestibulum* was then illuminated with an infrared LED (Francescato et al. 2002) (Fig. 6.45).

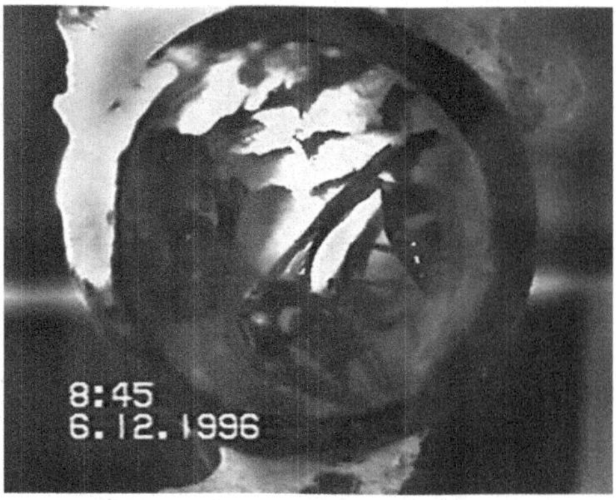

Fig. 6.45 A female of *E. fraterna* inside the vestibulum of her nest recorded by a micro video camera with the lens applied to an infrared filter closing a hole made in the external walls

The construction material is vegetable matter and the resulting paper is probably the most resistant of all the hover wasps. In some species (such as *E. fraterna*—personal observation— and *E. eximia*—Krombein 1991) nest material can be reinforced by fungal hyphae.

Here I report the characteristics of the nests of the most common species.

6.2.7.1 *Eustenogaster fraterna*

This is a mountain species, living in areas above 1,000 m and was studied by our group mainly at Bukit Fraser (Fraser Hill, Pahang State, Peninsular Malaysia). The nest is remarkably well camouflaged, resembling a dried fruit or leaf (Fig. 6.46), and is usually implanted on twigs, rootlets or thin leaves situated under embankments along roads or trails in deep forest (Turillazzi and Gerace 1992). The substratum does not penetrate the cells and an evident extra-alveolar reinforcement strengthens the nest attachment to the substratum itself.

As in all other known species of *Eustenogaster*, the nest consists of a comb of cells, rarely exceeding 20 in number. The *vestibulum* is formed by the prolongation of the peripheral cell walls which continue to form a tube (nest entrance) so that the entire structure has the appearance of an upside-down flask. Additionally, in this species, the lower part of the nest entrance is modified with a leaf-like, pointed, process and contains several small holes, probably allowing the wasps to see outside when guarding the nest entrance (Fig. 6.47).

At the cell comb level, the nest can present one or more "wings" which together with the pointed process and the grey coloration confer on the entire structure the appearance of a dead, crumpled leaf. These nests are the largest of all in the genus. I found one nest with 11 adults (five females and six males) all present at the same time.

The nest may possess a small gelatinous ant guard placed on the substrate 1–2 cm above the upper portion.

Fig. 6.46 A nest of *E. fraterna* hanging from a rootlet under an earth trench at Bukit Fraser, Peninsular Malaysia. A viscous, whitish ant guard is visible along the substratum (*arrow*) (photo by S. Bambi)

6.2.7.2 *Eustenogaster gibbosa*

Two nests of this species have recently been found and described in the area around the Gombak Field Station, 25 km NE of Kuala Lumpur. Maybe this is a rare species but it is also true that its nest is extremely difficult to find because it perfectly mimics a dead leaf (Fig. 6.48) (Hashim et al. 2006).

The outstanding peculiarity of the nest of *E. gibbosa* is the elaboration of the spout. There is a pair of very flat, broad flanges, one on each side, standing out from the main body of the nest and joining into a sharp point below the nest entrance. These flanges resemble the fins of a squid. An additional elaboration is a set of very

Fig. 6.47 Particular of nest entrance tube in a nest of *E. fraterna* with a female positioned on it (photo by D. Baracchi)

low ridges (eight in nest 1, nine in nest 2) running down the pseudoenvelope at an angle from the level of the comb.

In *E. gibbosa*, the ridges are arranged in a "twirl" rather than in generally "top-bottom" vertical lines as found in *E. fraterna*, *E. micans*, *E. calyptodoma* and to some extent *E. hauxwellii*. The colour of the vegetable material used for the construction is quite pale in the two nests described, both of which were found in wads of a parasitic horse hair fungus hanging from dead tree branches. It seems that the maximum size of these nests is rather smaller than *E. fraterna*, the species with similar nest architecture.

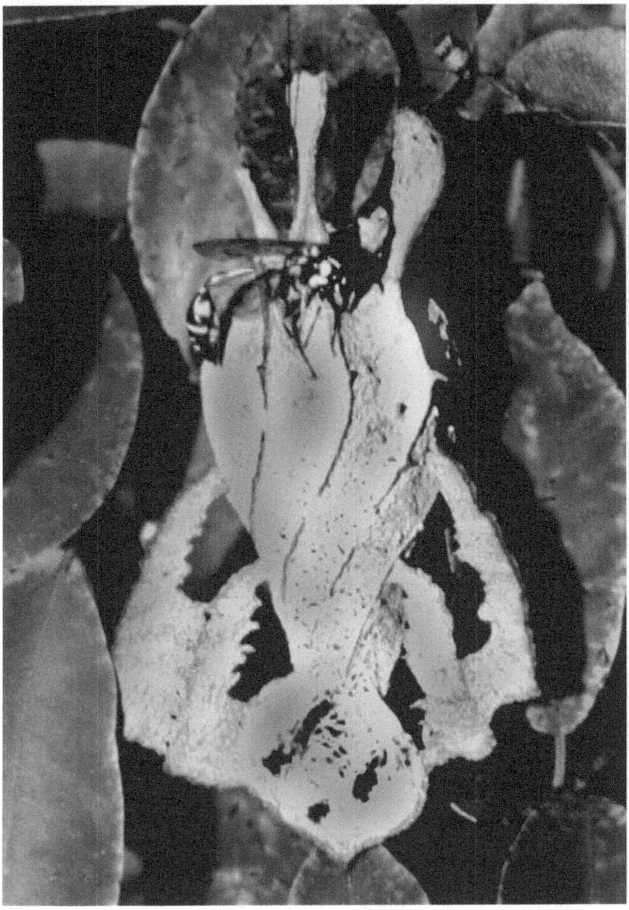

Fig. 6.48 A nest of *Eustenogaster gibbosa* with a female walking over the external walls (photo by Rosli Hashim)

6.2.7.3 *Eustenogaster hauxwellii*

This is an interesting species both regards its behaviour and its nest architecture. Indeed, this is the only hover wasp I have found to react in an aggressive manner similar to that of other social wasps when its colony is lightly disturbed. Females of a colony of 3–4 individuals attacked me when I was approaching their nest with the intention of photographing it. On several occasions, one female flew onto my camera lens in an attempt to sting it and then returned straight to her nest. The colony was the largest I had ever seen for this species and full of immature brood: I was never able to find another colony of that size so my observation remains the only one of this type.

Fig. 6.49 A nest of *E. hauxwellii* (photo by D. Baracchi)

The nest architecture is characteristic but still within the range of the overall design of the genus (Fig. 6.49). The *vestibulum* walls are built like lace in the lower part and there is no true entrance tube, so the nest lacks the characteristic inverted flask shape which we find in the other species. However, the nests are always defended by an ant guard. A good drawing of a nest which definitely belongs to this species is given by Ohgushi et al. (1983, p. 38, Fig. 31)

6.2.7.4 *Eustenogaster eximia*

The nest of *E. eximia* from Sri Lanka has been studied and described in detail by Karl Krombein (1991). He found nests of this species usually beneath earth or rock overhangs along streams in the rain forest attached to exposed roots or plant stems in a situation quite similar to the one I observed in *E. fraterna*. The substrate was

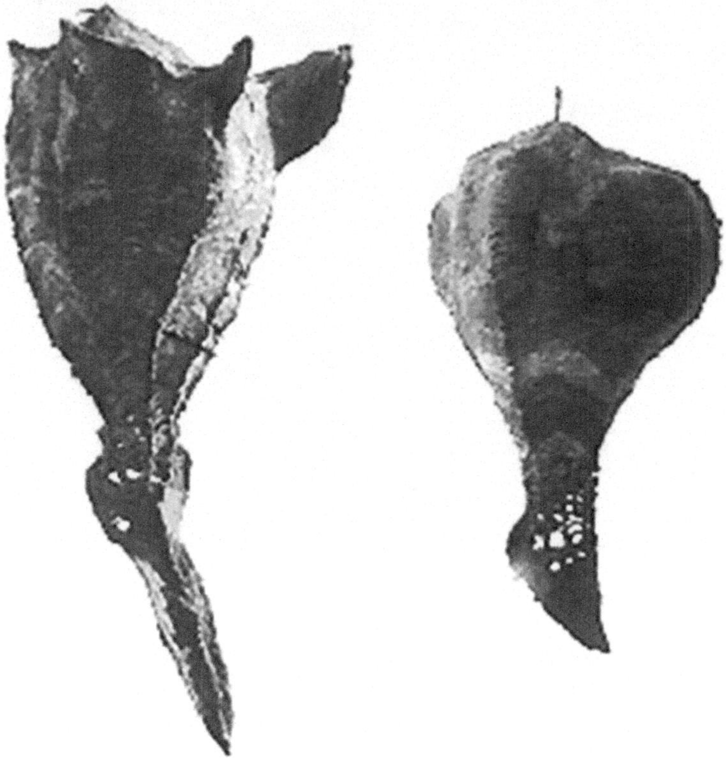

Fig. 6.50 Two nests of *E. eximia* studied by Krombein (from Krombein 1991 re-drawn)

encircled by the top of the comb but did not pass through any of the cells. The nest formed a structure with a 0.24–0.36 mm thick carton and the complete artefact was more or less pear shaped but quite variable, depending on the number and arrangement of cells which ranged, in the observed nests, from 4 to 26. Below the cells, organised in a single comb, the nest narrows to form a "pseudoenvelope" ending with a spout with the entrance at the end (Fig. 6.50). The author observes how the walls of the spout are perforated by a "lacy network of cartoon through which the wasp may peer". He also observes how the wide variation in the shape of the nests casts some doubts on the architectural categories of the *Eustenogaster* nests proposed by Ohgushi et al. (1983) stressing that "there may be considerable plasticity in nest architecture within a single species of *Eustenogaster*" and, I would add, in other genera too. To a certain extent this limits the concept that any species of hover wasps can be unquestionably identified by the architecture of its nest alone.

Krombein never found ant guards in this species, but he did find a nest with two spouts, each with its own viable entrance hole. This phenomenon is not all that rare in *Eustenogaster* and I have also found it in other species, such as *E. calyptodoma* and *E. micans* (personal observation). Krombein explains his finding as a "rare example of spout replacement following supersedure of one female by another".

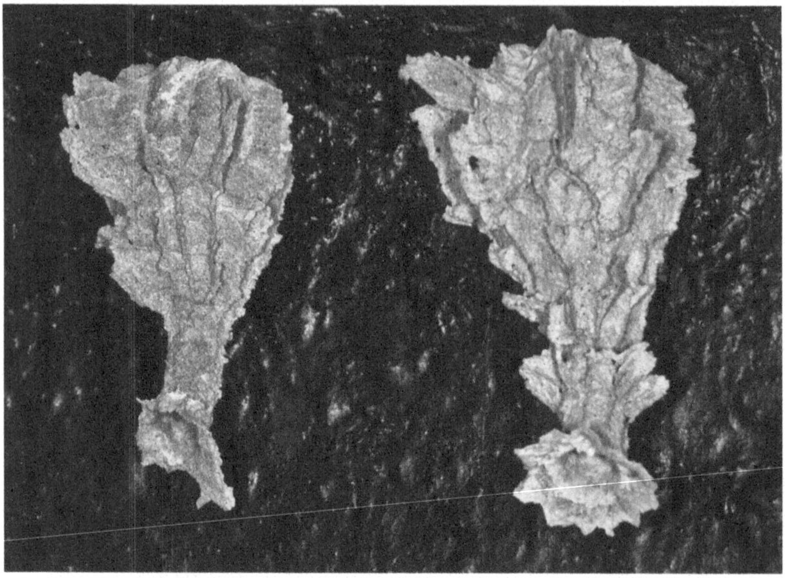

Fig. 6.51 Nest of *E. micans* (photo by D. Baracchi)

Another possible explanation interprets double spouts as a defence strategy aimed at fooling possible usurpers on entering the nest: this could be supported by the fact that in some cases only one spout presents a viable entrance hole (Fig. 6.55).

6.2.7.5 *Eustenogaster nigra*

Eustenogaster nigra is a species recently described by Fuki Saito and Lien P.T. Nguyen (Saito et al. 2006). This species is distributed from central Vietnam to South China in areas with more or less distinct seasons; males and females have been found overwintering inside their nests. Nest architecture is very similar to that of *Eustenogaster fraterna* and *E. eximia* from which it differs mainly for the longer entrance tube and for the slightly spiral envelope keels. Along the thread-like substratum of one of the two nests studied, Saito and colleagues found a curious tube-like carton structure which the authors interpreted as an ant guard or a drop stopper. Should this finding be confirmed through further observations, this structure will be a peculiar annex of the nest of this species.

6.2.7.6 *Eustenogaster micans*

Nests of this species can be easily found on flat substrates offered by rocks, large tree trunks and even human buildings. The silhouette is that of a slender flask, the neck of which ends with an opening that resembles that of a bottle (Fig. 6.51). This means that the number of cells which constitutes the comb is always limited to no

Fig. 6.52 A nest of *E. calyptodoma* on a building wall (photo by D. Baracchi)

more than a dozen. The *vestibulum* is quite restricted and colony size rarely exceeds two adult individuals. The external walls, however, are not smooth like a bottle but present longitudinal ridges. Nests potentially assigned to this species also exhibit structures attached to thread-like substrata (Williams 1919).

6.2.7.7 *Eustenogaster calyptodoma*

This is a relatively common and noteworthy species. Like the previous one, it builds its nests on flat substrates and quite often on buildings (Fig. 6.52). It is not uncommon to see these nests under the roof slopes of barracks and gazebos or other buildings, and they are impressive for their characteristic architectural design.

Fig. 6.53 The envelope in a nest of *E. calyptodoma* (here in phase of construction) is completely independent of the comb (photo by D. Baracchi)

As stated above, this is the only species in the genus (and in all the hover wasps) that constructs a true envelope which protects the comb of cells but which is completely independent from the latter. This may seem to be a very banal distinction but it is not. In fact the envelope is developed into a structure which is completely separate from the part of the nest devoted to brood rearing (Fig. 6.53). The evolutionary development of this sort of structure away from the simple walls of a *vestibulum* can be reconstructed as the progressive reduction of the most peripheral comb cells. However, the building sequence of the nest clearly consists in the development of two separate structures: the comb and the envelope (Fig. 6.54). The first cells are attached directly onto the substrate by their base or side, depending on the orientation (horizontal or vertical) of the substrate itself. The cell openings always face downwards.

The distance from comb to envelope is constant in the upper and lateral portion of the nest and is sufficient to provide passage to the wasps. The space, corresponding to the *vestibulum* found in other species is larger but the maximum number of cells is lower for similar-sized nests compared with all the other *Eustenogaster* species. The envelope entrance presents a characteristic crown formed by a varying number of spikes of material; this crown varies in aperture size and its functions are not yet clear. It is not uncommon to find nests with two entrance tubes (Fig. 6.55) one of which may be closed. We do not yet know if this is simply a mistake in the construction programme of the wasp or if it serves a true defence function (but see the explanation of Krombein 1991, in Sect. 6.2.7.4).

As in all the other *Eustenogaster*, the females of this species produce sounds by beating their abdomen on the internal walls of the *vestibulum*. *E. calyptodoma* is particularly "skilful" in doing this, but we shall look into this more in the paragraph on the defensive function of nest architecture.

Fig. 6.54 Construction of the envelope of a nest of *E. calyptodoma* (photo by D. Baracchi)

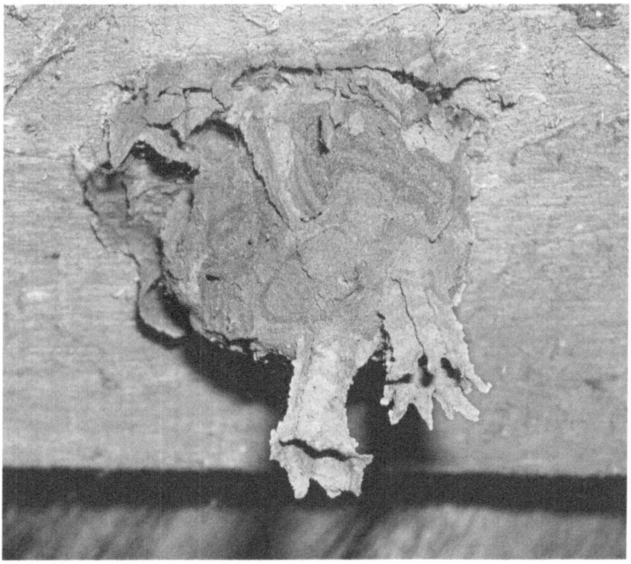

Fig. 6.55 A *E. calyptodoma* nest with two entrance tubes (one of which is blocked). Note the crown spouts around the tube openings (photo by D. Baracchi)

Fig. 6.56 A nest of *Metischnogaster drewseni* in a wad of *Marasmius* fungus (photo by D. Baracchi)

6.2.8 Metischnogaster

This genus includes two described species (*M. cilipennis* and *M. drewseni*) with the most highly camouflaged nests imaginable so that they are really difficult to find. They build their nests mostly on wads formed by a parasitic fungus (genus *Marasmius*) the hyphae of which resemble horse hairs. The fungi grow on dead trees or branches and the nests are easily confused with dead twigs wrapped by the hyphae. If one wants to find these nests it is usually easier to look for fungus wads and then to keep eyes peeled for the possible presence of nests. Sometimes the only hope of finding them is to look for the wasps, which normally rest out of sight inside a cell, patrol the nest or are actively involved in tending the larvae or building cells.

M. cilipennis nests and those of *M. drewseni* are quite similar. The first was briefly described by Pagden (1958, 1962). The second (Fig. 6.56) consists of a row

Fig. 6.57 The drop stopper of a *Metischnogaster drewseni* nest

of cells, the first of which is attached with its bottom to the tip of a thread-like suspension. The second cell is attached to the lower end of the first, the third to that of the second and so on.

All the alveolar openings face outwards in different directions; the angle between the openings of two consecutive cells is on average about 140°. Pagden reports a maximum of 10 cells for *M. cilipennis* (Pagden 1962) and I found a maximum of 17 cells for *M. drewseni*.

A special conical structure (which I shall call the "drop-stopper"), made of the same material as the nest, lies along the suspension thread about 10 mm above the first cell. These structures are not built to prevent small predators from getting at the nest cells, since ants can easily climb over them; rather they are probably devices to stop drops of water sliding down the structure (Fig. 6.57).On various occasions I have observed that while the drop-stopper over the nest was drenched with water after heavy rain fall and there was a drop of water inside the cone itself, the cells still remained dry. Some nests may have more than one cone-like drop-stopper.

The whole nest perfectly mimics a dry twig in shape and colour. The material used for the construction is mainly vegetable matter and the whole structure is very lightweight.

Notwithstanding the poor quality of nest material and the fragility of the cell walls, the nest structure is flex resistant and elastic to a certain degree. During and after rainstorms, the nest of a colony I was observing was often soaked with rain dripping down the suspension thread, which the drop-stopper only managed to block in part. I twice observed that when the nest (about 10 cm in length) quickly dried out in brilliant sunshine, it bowed to form an arc of about 120° and then resumed its original shape after a few minutes. The wasps did not seem at all perturbed by these changes and continued to rest inside the cells without any apparent signs of reaction (Turillazzi 1990b).

6.3 The Nest as a Defence for the Immature Brood

6.3.1 Nidification Sites

The main predators of the Stenogastrinae brood, as in other social wasps, are ants (Fig. 6.58) (Jeanne 1975, 1979; Samuel 1987; Turillazzi 1991) and hornets of the genus *Vespa* (especially *V. tropica* (Hansell 1982; Matsuura 1991; Gadagkar 1991) (Fig. 6.59). Their immature brood may be a tempting morsel for small vertebrates too and even bats in caves in the Philippines have been reported (Chris Starr, personal communication) preying on nests of *Parischnogaster*. Indeed, they have evolved low active defence mechanisms against larger predators and, living in small colonies (3–4 individuals), they cannot rely on recruitment of forces for defence. The choice of a good nesting site can be important for the success of the colony and this explains why the nests of some species are found in places where ants are not so common, such as the slippery walls of caves or overhanging rocks or in dark places where hornets cannot fly or detect the nests. In other cases implanting a nest at the top of a thin suspension, such as a twig or a rootlet or on a leaf in a forest can greatly reduce the probability of being found by a small predator searching at random.

6.3.2 Nest Camouflage

Nest camouflage is the main defence against predators which use visual cues. As we saw in the descriptions of the nest architecture in the various species, this confers

Fig. 6.58 Ants attacking the nest of *Eustenogaster calyptodoma* (photo by D. Baracchi)

the structure a set of characteristics which are probably the result of selective pressure exerted by the fiercest predators of these wasps, hornets. In the South Asian forests, hornets curb colony development of all the other social wasps, and the same tendency to build mimetic nests of limited size is found in species of the genera *Ropalidia*, *Polistes* and *Parapolybia*. Whilst the low quality of construction material is one key factor that prevents the establishment of a high level of social organisation in the hover wasps, it is also certainly true that this latter important evolutionary pressure has also limited their colony size. However, the various genera and species have achieved perfect camouflage in different ways with an incredible variety of results.

Looking at the nests and forest vegetation it is obvious that mimicking plant parts or other forms of camouflage is a common tactic in hover wasp nest design (Fig. 6.60). Christina Coster-Longman reports various nest-mimic pairs, summarised in a table in her, in part unpublished, PhD thesis (Coster-Longman 1998) (Table 6.1).

6.3.3 Ant Guards

We define ant guards as gelatinous or sticky barriers, placed to defend the approach to the nest, as particular structural features of nest architecture. As we saw, these structures occur in the nests of at least three genera (*Eustenogaster*, *Liostenogaster* and *Parischnogaster*) and possibly in *Stenogaster fulgipennis*, according to the description by Lieftinck reported by van der Vecht (1975). In all cases they are built

Fig. 6.59 *Vespa tropica* attacking a *P. striatula* nest (photo by D. Baracchi)

using the gelatinous secretion produced by the abdominal Dufour's gland (Sledge et al. 2000). In *Eustenogaster* we find only some species which use this defensive resource (Fig. 6.46): in *E. fraterna* in a sample of 15 complete, active nests, I found only two with fresh ant guards on the nest substrate. However, we found fresh ant guards in at least three out of eight nests where the envelopes were still incomplete. In *Eustenogaster hauxwellii*, all the nests that I found had fresh ant guards, including the most mature nests, and sometimes the ant guards had been applied to the substrate on either side of the nest (Fig. 6.61). During application of the ant guard, the wasps cling to the substrate with tarsi held close together, wings closed and the head held downwards. The mandibles may clasp the substrate or remain closed, while the antennae are held immobile. The peduncle of the wasp is held parallel to the substrate and the gaster is curved towards the substrate and pointed towards the head (Fig. 6.62). The terminal parts of the gaster are then stroked backwards and forwards over the substrate, covering an area of approximately 1–2 cm. We measured the frequency of the to-and-fro movements by a single

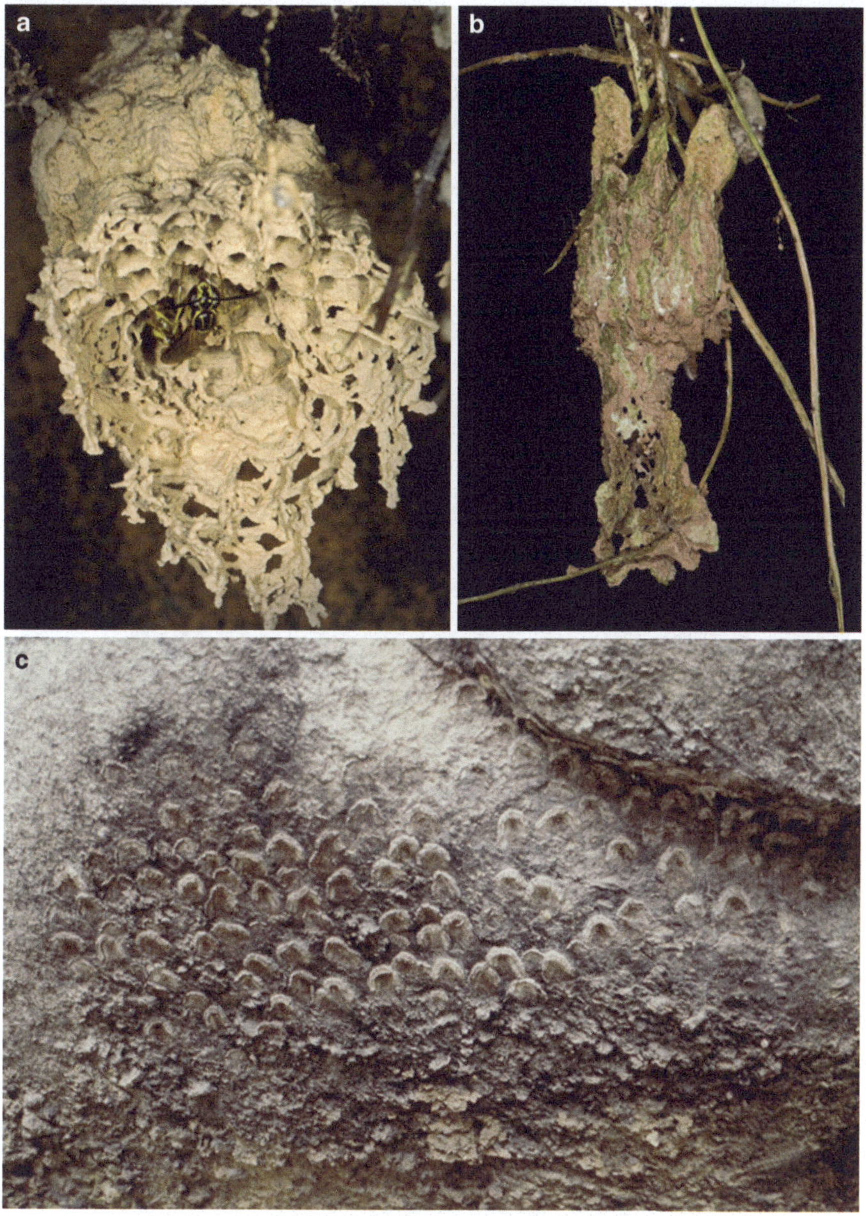

Fig. 6.60 Examples of nest camouflage in Hover Wasps. (**a**) *Liostenogaster flavolineata*; (**b**) *Eustenogaster* sp. (photo by D. Baracchi); (**c**) nest cluster of *Liostenogaster vechti*

female as approximately 57 in 30 s. At the same time, the wasp revolves around the substrate with subtle movements of the legs, completing a revolution within 40 s. During this application, a thin film of grey–white secretion can be seen extending from the abdomen to the secretion already applied to the substrate.

Table 6.1 Some stenogastrine nests and mimicked natural objects observed in the field (from Coster-Longman 1998)

Species	Match
Liostenogaster vechti	Lichens, tree bracts
L. flavolineata	Mud balls
L. topographica	Roots, termite runs
Parischnogaster mellyi	Dried vine leaves
P. jacobsoni	Thorns, fruits of *Callistemon* spp.
P. striatula	Thick twigs/stems
Metischnogaster sp.	Dead twigs
Eustenogaster spp.	Dried fern fronds and flowers

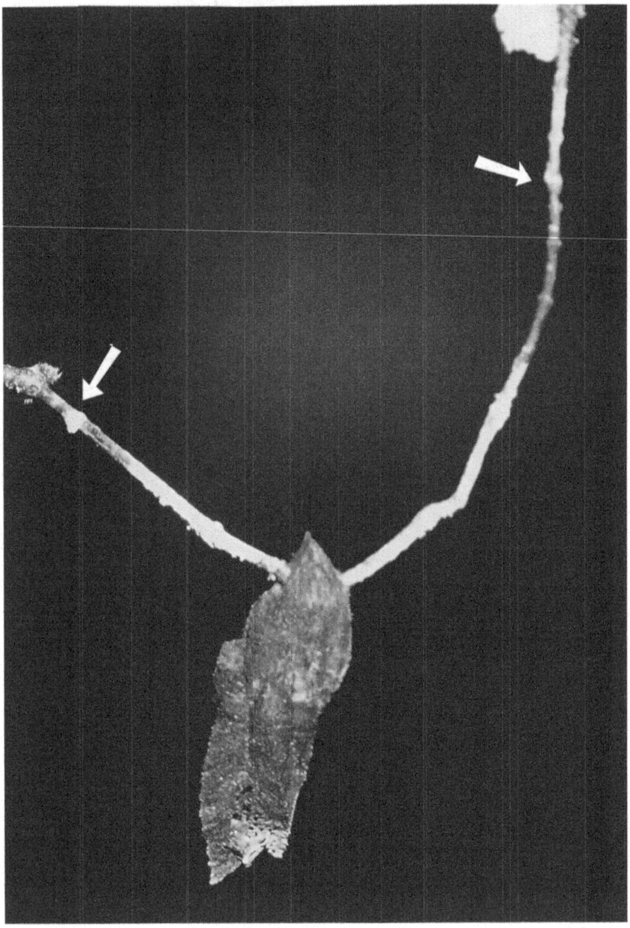

Fig. 6.61 The ant guards (*arrows*) of a nest of *E. hauxwellii* (photo by S. Bambi)

A recently described species from China, *E. nigra*, which constructs a nest quite similar to that of *fraterna*, also appears to have ant guards (Barthélémy on line). In general, we can say that only certain species which build nests on thread-like

Fig. 6.62 A female of *E. hauxwelli* applying the Dufour's gland secretion to form the ant guard of her initial nest

substrates fashion ant guards; but it cannot be excluded that only some individuals exhibit such behaviour or indeed that the nest is supplied with this defence only at a very early stage of construction.

The only species of the genus *Liostenogaster* which builds ant guards is *Liostenogaster tutua*. As we have already seen, this species usually constructs nests with muddy material and places them at the end of horizontally oriented twigs and small branches. The ant guard is deposited approximately 2–4 cm away from the first cell and construction behaviour is quite similar to that described for *Eustenogaster*. This results in the application of the secretion over a large surface area of the substrate, both by direct and indirect application, as a result of the gaster further spreading previously deposited secretion (Fig. 6.63).

The behaviour which leads to the construction of the ant guard in some species of the genus *Parischnogaster* (especially those belonging to the *jacobsoni* group) is, on the contrary, completely different; while in the other genera the ant guard is simply a patch of secretion smeared on the nest substrate, in this genus the ant guard has a definite architectural structure. If we look at ant guard construction behaviour in *P. nigricans serrei*, we can observe that it begins 1–2 days after the construction of the first cell. Tiny drops of the Dufour's gland secretion are collected on the hind legs by rubbing them against the extremity of the slightly flexed abdomen and then

Fig. 6.63 The ant guard construction in a nest of *Liostenogaster tutua*

passed in rapid succession via the middle legs into the mouth of the wasp. She applies this by repeatedly licking either the support or mounds of previously applied substance, simultaneously moving around the support without ever losing contact (Fig. 6.64). The entire sequence can last for as long as 30 min and is usually performed in the late afternoon just before sunset.

In new nests, the structure consists of a single ring about 1 mm high but in mature nests it acquires various shapes from a pad with small rounded knobs to disk-like ridges (Fig. 6.65) (Turillazzi and Pardi 1981). I have never found a single example of ant guard in hundreds of nests of *P. mellyi* that I have had the chance to observe during my research but recently I found traces of freshly deposited ant guard on the substrate of a single cell foundation built by a female of this species. This suggests that this kind of defence could also be used in some species of *Parischnogaster* by some individuals only and then at a very early phase of nest construction.

In nests of the species which use the flat substrates of leaves to place their cells, the ant guard is invariably deposited on the leaf petiole. These ant guards proved to be effective in repelling or blocking the passage of ants, at least the smaller species (Turillazzi and Pardi 1981). The chemical composition of the secretion is mainly based on long chain hydrocarbons but is specific to each species and, as stated above, is quite similar to the substance used for rearing the larvae (Sledge et al. 2000) (Fig. 6.66). The same authors also found that the Dufour's gland secretion of *E. hauxwellii*, whose nests always feature ant guards, is more complex (36 compounds) than that of *E. fraterna* (16 compounds) which exhibits ant guards almost exclusively on very young nests. Moreover, *L. tutua* (54 compounds), *Parischnogaster* sp. and *P. jacobsoni* also produce richer secretions than species that do not produce ant guards, such as *L. flavolineata* (Keegans et al. 1993) and *P. mellyi* (only 14 compounds). This suggests that the particular contents of the

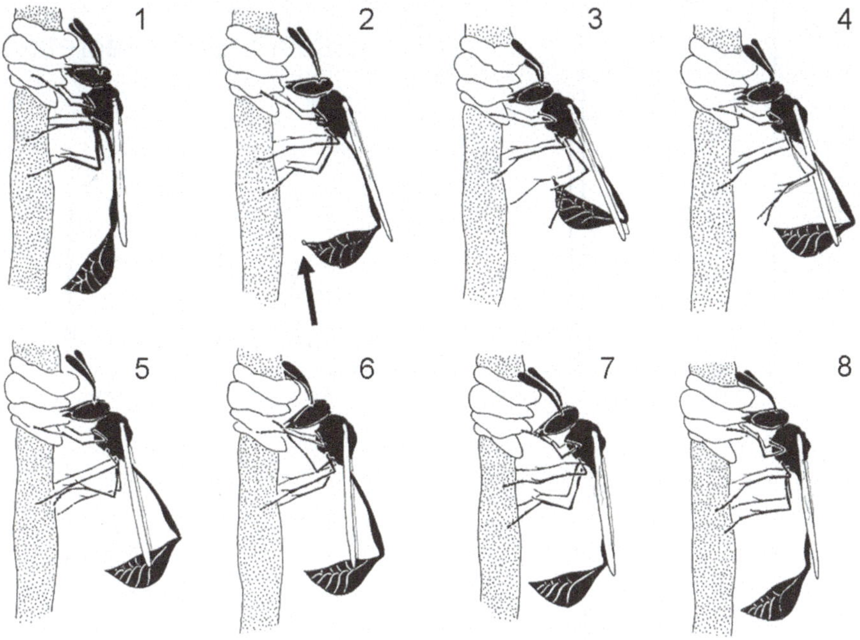

Fig. 6.64 Sequence of ant guard construction in *Parischnogaster nigricans serrei* (from Turillazzi and Pardi 1981, re-drawn). Note the small drop of Dufour's gland substance emitted at the tip of the gaster (2), the movement of the legs to bring it to the mouth (3–7) and its application to the previous ant guard material (8)

Fig. 6.65 Various ant guards on nests of *Parischnogaster jacobsoni* and *P. nigricans* (photo by D. Baracchi)

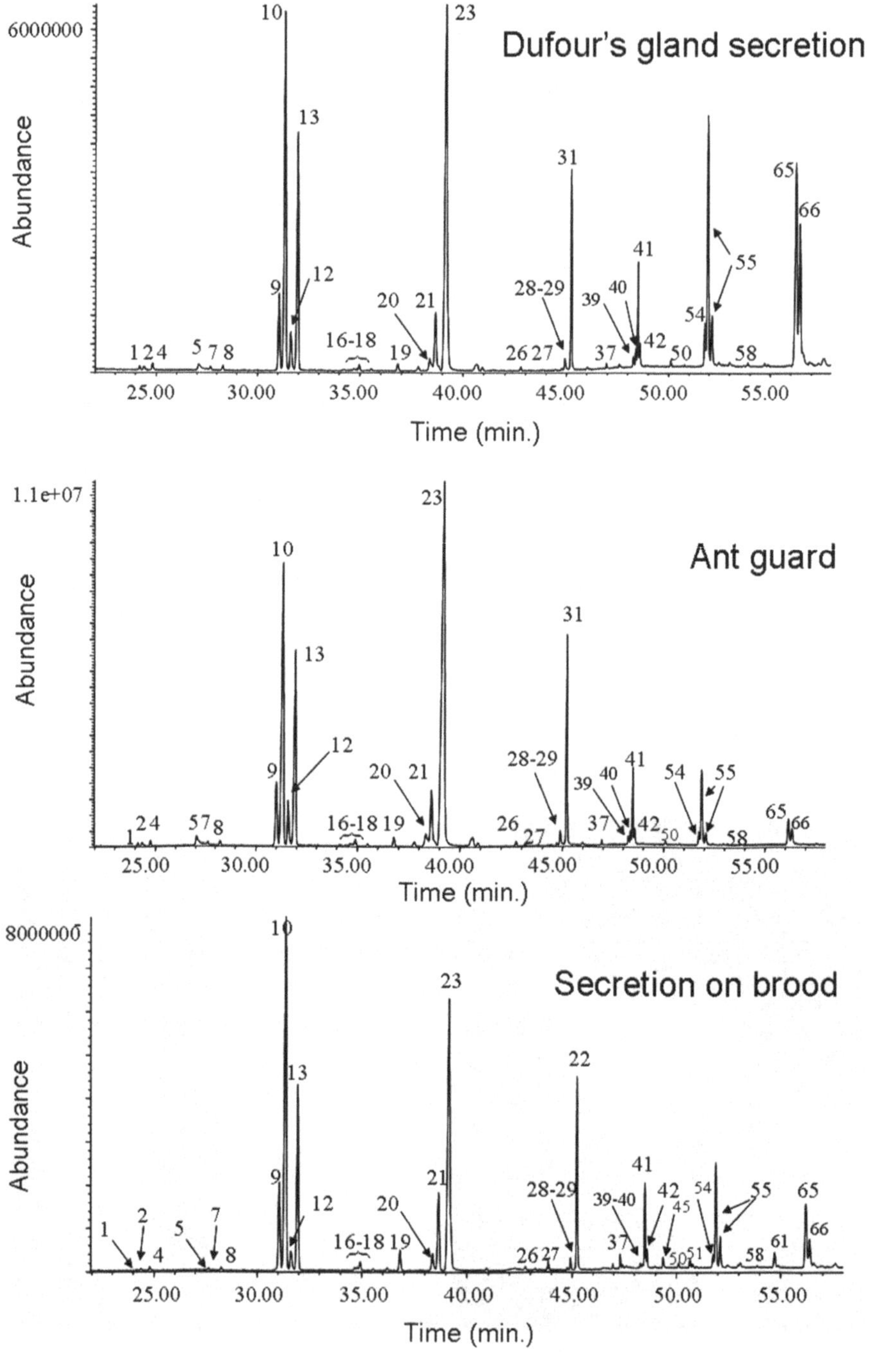

Fig. 6.66 Total ion Gas Chromatograms of Dufour's gland secretion, ant guard secretion and brood rearing secretion of *Parischnogaster* sp. sim *jacobsoni*. Same numbers correspond to same compounds (modified from Sledge et al. 2000)

Dufour's gland of the various species may have evolved as a result of selection for the production of an efficient ant guard for nest defence. This function could have arisen through an increased number and variety of substances in species that do produce ant guards, compared with those that do not.

Species which use this sort of defence usually place their cells on thin sticks, branches, roots or wires, but it does seem to be adopted by those species which nest on even thinner substrates, such as *Metischnogaster* sp. or *Eustenogaster gibbosa*.

It is unclear whether species which nest on flat substrata, such as *Liostenogaster vechti* or *Liostenogaster topographica*, also use chemical defence against ants. My student Christina Coster-Longman performed several experiments on nests of both species. She found that ants of an undetermined species passing in columns near the nest area were not stopped by a mixture of *L. vechti* nest-patch material and water. Hexane extracts of *L. vechti* nest-patch material tested on the European ant *Crematogaster scutellaris*, however, did produce some positive results with respect to pure solvent (Coster-Longman unpublished data). With regard to *L. topographica*, we recently observed that ants are prevented from entering the nest area by very tiny particles of material that the wasps apply to the substrate around the cells. Ants seem to avoid walking on this material, just as they do when they find the surface covered with talcum powder (an expedient often used to prevent them escaping from laboratory cages). Hexane extracts of the same material have no repellent effect. Thus, this last species seems to rely on a physical barrier rather than a chemical one for defending its nests from ant attacks (Baracchi et al. 2009).

6.3.4 Limiting Access to the Cells

One method of defending the cells and the brood inside is to limit access of potential predators with special architectural structures. These can be of various types but the simplest is evidently the solution found by *Liostenogaster abstrusa* which, as we have seen, nests directly in dead hollow branches, thus providing its nests excellent protection against visual hunting predators such as hornets. Other solutions are, instead, clearly architectural. If we maintain a structural, but not functional, distinction within the Stenogastrinae, then only the nest of *Eustenogaster calyptodoma* possesses a true envelope intended as a totally independent structure from the nest itself. The nests of all the other species of the genus *Eustenogaster*, together with the flask-shaped mud nest found in New Guinea and attributed to an unknown species of *Stenogaster* by Hansell and Turillazzi (1991), present instead what has been defined a "pseudo-envelope" (although Wenzel 1991 p. 499, observes that the term is incorrect as both structures—the envelope and the pseudo-envelope—"may have been originally derived from extensions of the cells wall"). Another solution (clearly derived from nests similar to that of *L. flavolineata*) is reached by *L. pardii*—the cells in its globular nest all open facing the centre of the globe, thus defining a common *vestibulum* with a single, restricted, entrance (Fig. 6.67).

Fig. 6.67 A female of *Liostenogaster pardii* guarding the entrance of her nest. (Photo by D. Baracchi)

In the *Parischnogaster striatula* group (including *P. alternata*), the architectural solution has been to build the cells, arranged in linear or twisted rows, with their openings facing a common tube. In this way external access to the cells is limited to only one entrance hole. These nests restrict the movements of the adults when inside the tube, far more so than in the other two kinds of enveloped structures. In conclusion, we have three cases where the architecture of the nests has been admirably adapted to counteract the selective pressure of predators, limiting access to the cells and brood inside them to a single hole, which can be more easily defended by a single wasp. These are evidently adaptations that species belonging to three different genera have acquired independently over the course of evolution.

6.3.5 Nest Clustering

One defensive strategy often used by various animals is to aggregate in great numbers, thus forming large groups of individuals to challenge predator aggression or dilute the effect of predation. Some stenogastrine species tend to build their nests in particular places forming clusters of colonies that in some cases can reach incredible sizes. This is not directly influenced by the architectural characteristics of the nests and the phenomenon is found in three species in particular (*Liostenogaster vechti*, *L. flavolineata*, and *Parischnogaster alternata*) which build diversely shaped nests of different construction materials but, mainly, on flat substrata (Fig. 6.68). It is clear from data collected by researchers of my group, and in particular by Christina Coster-Longman, that the three species are choosy in

Fig. 6.68 A huge cluster of *Liostenogaster flavolineata* nests (the pale, massive ones) and *L. vechti* (the greyish, flat ones) under the culvert of a bridge

selecting their nesting site. Light seems to play an important role in determining nest location: *L. vechti* always chooses the most illuminated walls and ceiling of buildings or grottoes. On the contrary *P. alternata* seems to prefer darker sites, as seen by various clusters of this species found in pipe tubes, on the ceiling of caves and the interior of abandoned huts. *L. flavolineata* appears more flexible, being found associated with both species in some clusters. Other important factors could be the availability of good materials for the construction of the nests, such as spider silk for *P. alternata* and vegetation rich in phenols and terpenes for *L. vechti*.

There may be several mechanisms which determine this phenomenon; the most probable being the tendency of young foundresses to establish their nest near the maternal one. Re-colonisation of the Bukit Fraser gazeboes by *L. vechti*, after the quasi-annual destruction due to refurbishing works, is further evidence of the philopatry of the wasps to return to a well-known site. Whatever the case, it brings important consequences for the success and social life of the single colonies, including protection from predators.

Probably the worst risk for species nesting in clusters is the higher possibility of predator attack. Nest clusters are easily visible on account of their poor camouflage. The risk a cluster runs of catching the attention of a predator should be balanced by the benefit individuals gain from group living, i.e. indirect defence due to the dilution effect (Turner and Pitcher 1986; Irman and Krebs 1988). Experiments conducted on clusters of colonies of *P. alternata* brought interesting results. In a cluster, the central core of older nests is surrounded by younger colonies. This testifies that cluster geometry is not random but plays an important role in the development of the cluster over time. Areas of higher nest density exhibit lower

Fig. 6.69 Workers of *Vespa tropica* (arrows) attacking *Liostenogaster vechti* colonies under a small bridge

V. tropica predation rate and central colonies are more protected than external ones, as predicted by the selfish-herd theory. Selfish-herd protection is also effective against ichneumon flies. These enemies were never seen to venture further than the first few peripheral nests in the agglomerate. Maybe this protection is also due to the particular nest architecture of this species. Nests hang down vertically from the ceiling, forming a sort of screen which may block access to flying intruders trying to reach the centre of the agglomerate (Landi et al. 2002).

Different results were obtained for *Liostenogaster vechti* (Coster-Longman et al. 2002). Central colonies gain protection from more external colonies against ant attack; however, the selfish-herd effect does not seem to be effective against ichneumon flies and *V. tropica* (Fig. 6.69). Probably these two last-mentioned enemies can reach the centre of the cluster because the flat shape of the nests does not obstruct their passage.

Our observations also revealed that different active defence strategies are employed against different predators. Whilst ichneumon flies are driven away by the residents of attacked colonies, preventing them from reaching central nests, *V. tropica* causes the hover wasps to fall from their nests. This behaviour saves adult individuals from the predator and, only occasionally, do they face the hornet, even if such attempts are of little effect.

In *L. vechti* new nests were always preceded by the application of a foundation "patch" and the wasps can spend up to 5 weeks on this preparation stage before constructing cells. This part of the nest could be considered as a sort of area of occupancy for testing the safety characteristics of the nesting position before investing in the nest proper. Moreover, some observations suggest that agglomerations are formed by groups of individuals at the same time, rather than by a single pioneer joined by others later, which could invite further colonisation by other wasps (Coster-Longman 1998).

Thus we have seen how, through nest site choice, nest architecture, nest camouflage and mimicry all three species cope excellently with the question of defence.

6.4 Materials and Instruments

As Mike Hansell points out (Hansell 2008) nests are the results of behaviours applied to materials. As with the other social wasps, hover wasps do not dig out their nests but build structures based on simple modules, called cells, where they rear their immature brood through to complete larval development. They use both inorganic and organic materials and are able to collect them from the environment and to process them before their utilisation for the construction of the nest structure. Unlike most of the social bees, wasps do not secrete their own nest material, with the exception of the secretion which is added to collected and elaborated matter to increase the compactness and strength of the nest structure. Material collected from the environment is rarely applied to the building without modification; rather it is processed with the mouth parts to make it more suitable for construction. Processing may also involve bringing together different materials in order to form a composite which can combine the virtues of the various components. Even the progression of construction itself can influence the mechanical properties of the structure and Hansell (2008) observes how the addition of material along the growing edge of a brood cell or to the envelope determines particular architectural characteristics to the nest.

Materials such as mud or vegetal fibres have different properties: mud is particularly suitable to sustain weights in compression while the material called paper is good for manipulation into thin and light sheets which are strong in tension. Various methods have been used for the study of nest material composition in hover wasps. One method for evaluating the relative proportions of organic and inorganic components in nest material was devised by Hansell (1984) and consists in the incineration of nest fragments in a porcelain crucible over a Bunsen burner for some minutes. The crucible is first weighed on a 1/10,000 g precision micro-balance by itself, then with the nest intact and finally with the burnt fragments (ashes), so that the percent of material lost to the atmosphere (organic) can be calculated from the original weight of the nest fragments and that of the respective ashes. Other methods were employed by Hansell and Turillazzi (1995) to determine the nature of material used by species of *Anischnogaster* in nest construction, by examining under the light microscope pieces of the nest previously placed in "Histoclear" and then broken up mechanically into their constituent fragments. Intact nest material was also observed under the Scanning Electron Microscope to determine the nature of its components, as well as the secretion which lines and binds the cell walls and the fungal hyphae growing on them (see also Krombein 1991). As a third analysis, penetration of fungal hyphae into the cell walls was studied by examining stained histological sections under the light microscope (Krombein 1991; Hansell and Turillazzi 1995).

Table 6.2 Percentages of inorganic material and size characteristics of the cells of nests of various species of stenogastrine wasps (from Bongiovanni 1998)

Species	*N*	Average % of inorganic material	SD	Cell diameter	Cell length	Wall thickness
L. nitidipennis	6	1.75	0.09375	4.17	9.98	0.72
L. vechti	6	5.69	0.25625	4.18	10.23	0.65
P. jacobsoni	6	7.21	7.21	2.38	6.2	0.46
P. mellyi	6	8.67	4.52	3.67	10.5	0.39
E. fraterna	6	8.98	4.59	6.88	13.16	0.44
E. calyptodoma	6	13.23	0.1875	5.62	11.52	0.56
P. alternata pointed	6	13.76	?	3.48	9.31	0.48
P. alternata flat	6	31.25	?	4.14	11.83	0.51
P. striatula smooth	6	16.96	?	3.4	7.8	0.34
P. striatula rough	6	23.75	7.45	3.4	7.8	0.52
L. pardii	6	92.02	1.54	4.08	11.38	0.86
L. flavolineata	6	92.09	2.42	5.31	19.1	0.64
L. campanulae	6	92.12	2.49	4.59	12.38	1.05

The quantity of binding matrix in a nest can be roughly estimated on the basis of images taken at the SEM as *scarce*, when the particles of the material are not coated at all (for example when the tissues can be recognised in cases of structures built with vegetable material); *medium*, when the particles are coated by a thin film and their nature can be hardly recognised; *abundant*, when the entire surface of the nest is covered by a thick film which hides the underlying material (Bongiovanni 1998, see also Kudô et al. 1996).

Students in my research group performed a detailed analysis of the building material used by various species belonging to the major hover wasp genera. From this survey it was clear that only the genus *Liostenogaster* presents species with nests built almost entirely of mud (with a percentage of inorganic material over 90 %) and species with nests built mainly with organic material (with a percentage of inorganic material less than 8 %). In the genus *Eustenogaster*, percentages of inorganic material vary from 8.97 % in the comb cells of *E. fraterna* to a maximum of 29 % in the envelope of the same species. In the genus *Parischnogaster* we also found very low percentages of inorganic material in some species (7.21 % in *P. jacobsoni* and 8.76 in *P. mellyi*) reaching values of 31.25 % in nests of *P. alternata* implanted on flat substrata. In any case intraspecies variations can be important, for example, in *P. mellyi* we found a range spanning from 3.88 % to a maximum of 16.76 % ($N = 6$, $SD = 4.52$).

Table 6.2 gives the relative percent amounts of organic and inorganic materials determined in species of three genera of hover wasps using the incineration method.

Moreover, dividing the particles (mineral or organic) found in the nests of some of the same species according to their size, we obtained the percentages reported in Table 6.3.

Table 6.3 Particle size and quantity of binding matrix in the material of the nests of some species of stenogastrine wasps (from Bongiovanni 1998).

Species	<2 μm	2–100 μm	100–400 μm	Matrix
E. calyptodoma comb	0	64.9	32.5	Abund
E. calyptodoma envelope	0	9.7	84.7	Scarce
E. fraterna	0	62.3	35.4	Abund
L. nitidipennis	0	53.8	58.5	Scarce
L. vechti	0	77.4	27	Abund
P. alternata flat	0	76.3	20.5	Medium
P. alternata pointed	0	76.3	20.5	Medium
P. jacobsoni	0	71.6	20.7	Abund
P. mellyi	0	74.9	24.6	Medium
P. striatula rough	0	75.7	22.8	Medium
P. striatula smooth	0	75.7	22.8	Medium
L. pardii	12.03	54.3	33.3	–
L. campanulae	12.08	58.7	28.3	–
L. flavolineata	13.08	61.4	24.8	–

Hansell suggests that the builders can select mud according to its properties (water content and mineral or organic composition) and precisely for its suitability for the structure to be built: this seems the case in hover wasps, especially for the envelope of *E. calyptodoma* which is built with particles of larger size with respect to the cell comb (see the following paragraph).

Vegetable particles are usually very small and taken from rotting wood. Moreover the salivary glands do not seem to be large enough to produce a great quantity of secretion and, as a result, these wasps generally produce very fragile nests which are not covered by an additional layer of secretion, as found in various other social wasps. This finding is confirmed by Kudô et al. (1996) who ascertained that the amount of binding matrix in *E. calyptodoma* is actually far lower than that employed by polistine and vespine wasps. In various cases the incomplete silk cocoon spun by the pupating larvae can be used to reinforce the cell walls. In other cases (like those reported in species of the genus *Eustenogaster*—Krombein 1991— and *Anischnogaster*—Hansell and Turillazzi 1995), the wasps tend to use a wide variety of material, including parts of the skeletons of dead prey, silk, plant hairs, etc. which seem to facilitate or induce the growth of algae and fungal hyphae over the surface of their nests and thus reinforce their structure or achieve good camouflage. It has also been demonstrated that materials can differ in different parts of a nest, for example, in *L. vechti* Christina Coster-Longman found that the nest "patch" and nest proper are of different composition. She performed Gas Chromatographic analyses on the materials and found that the patch has a simpler composition with respect to the cells. Moreover, on burning fragments of the nest, Christina found that they gave off a very aromatic odour and from the gas mass analysis it appears that the wood particles contain aromatic substances such as terpenes, phenols and sesquiterpenes. She hypothesises (Coster-Longman 1998) that careful choice of this particular daubing material by the wasps is a system for

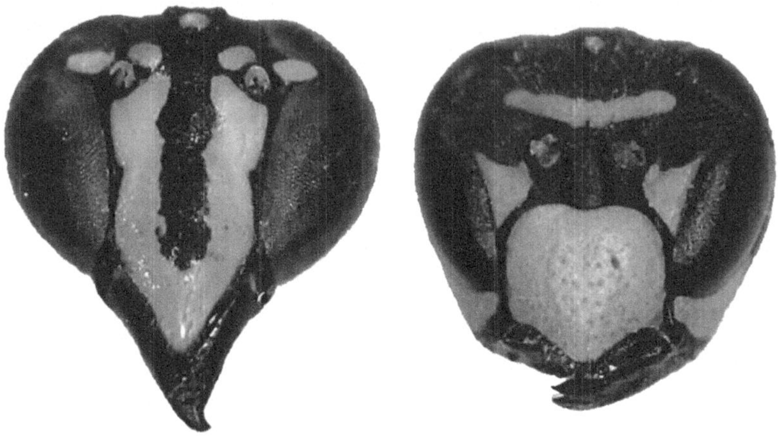

Fig. 6.70 Mandibles of a female *Eustenogaster* (*left*) and a female *Polistes* (*right*)

limiting ant attacks, since the aromatic substances could cover the odour of the brood. Monoterpenes and sesquiterpenes are known to be effective deterrents against ants and spiders (Eisner et al. 1974).

The construction material that we find in the nests of hover wasps is of clearly inferior in quality to that used by Polistinae and Vespinae and probably comparable to that of the Eumeninae. The paper of the walls in the nests of the Stenogastrinae is usually thicker than that found in similarly sized Polistinae nests. Long fibres of seasoned wood and good gluing capacities of the salivary secretion are necessary for the constitution of a good paper paste, but hover wasps are endowed with long and thin mandibles (and less developed mandibular muscles) which are not suitable for scraping material. In this respect Hansell (2008) observes how the angle formed between the mandibles when closed ranges from 45–70° in the Eumeninae to over 80° in the Polistinae, with hover wasps presenting intermediate values. To appreciate this, it is sufficient to compare the mandibles of a *Eustenogaster* with those of a *Polistes* (Fig. 6.70).

Some images obtained under SEM well illustrate the particular aspect of nest texture in different some species (Fig. 6.71).

One important distinction of *Liostenogaster* females is the shape of the mandible teeth which is pointed and blunt in some species and squared and sharp in others (Fig. 6.72). As already mentioned, this seems to be mainly connected with the material used to build the nest. This genus seems to be especially characterised by the fact that some of the species (*L. flavolineata*, *L. pardii* and *L. campanulae*) build heavy, massive mud nests while others (*L. vechti*, *L. nitidipennis*, *L. abstrusa*, *L. filicis*. and *L. topographica*) use plant materials. Hansell (1987b) has suggested that the return to the use of mud was a reaction to predatory pressure exerted on these wasps by tropical hornets and not a primitive character of the genus, as supposed by other authors (Ohgushi et al. 1990).

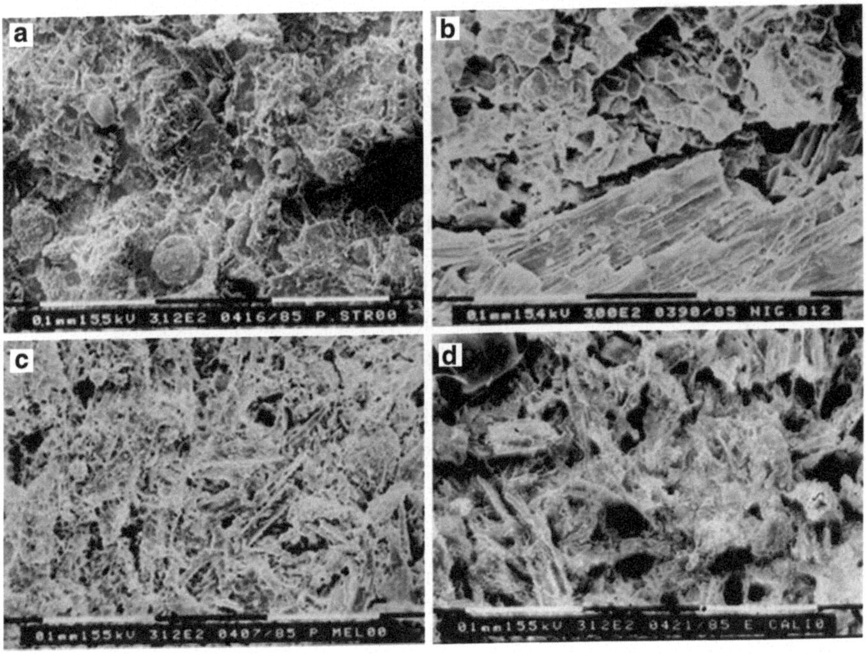

Fig. 6.71 Set of images obtained at the Scanning Electron Microscope of nests surfaces of some species of stenogastrine wasps (**a**) *Parischnogaster striatula*, (**b**) *P. nigricans*, (**c**) *P. mellyi*, (**d**) *Eustenogaster calyptodoma*

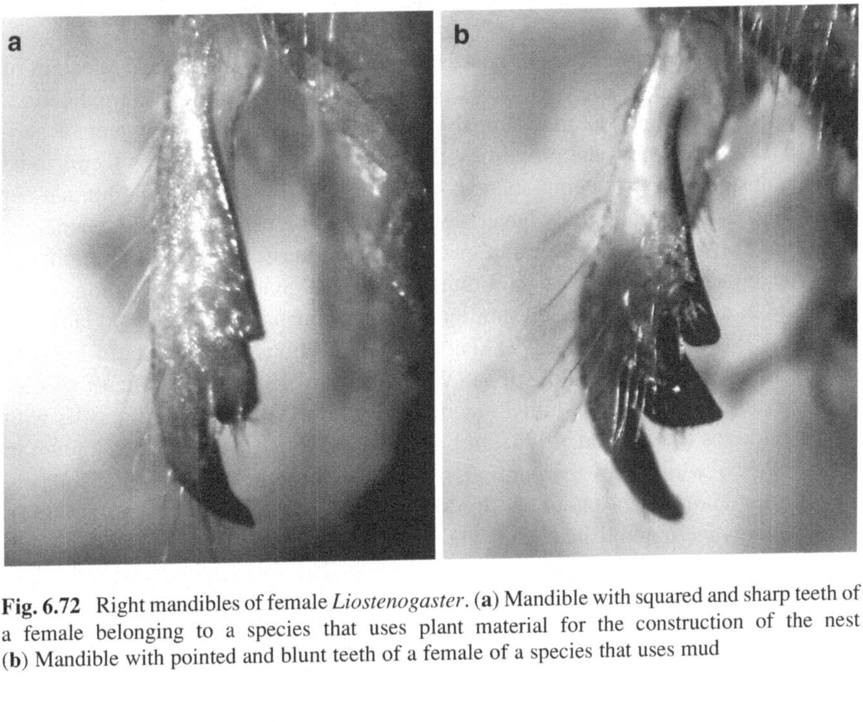

Fig. 6.72 Right mandibles of female *Liostenogaster*. (**a**) Mandible with squared and sharp teeth of a female belonging to a species that uses plant material for the construction of the nest (**b**) Mandible with pointed and blunt teeth of a female of a species that uses mud

6.5 The Nest as a Tool

Use of tools by insects has been observed in various cases: for example, solitary Sphecid wasps of the genus *Ammophyla* hold small pebbles in their mandibles which they use to press sand particles together to close the entrance of the cells they dig in soil; while ants have been often observed to place sand grains on sticky surfaces (such as tree resin) to reach food sources. The hover wasps offer two examples in which the nest structure itself is used as a sort of tool by its occupants.

Metischnogaster drewseni (Fig. 6.73) employs a particular form of active defence to get rid of small walking predators, such as ants, which may arrive on the nest. I had the opportunity of observing this on a number of occasions. It mainly consists in several attempts by the wasps to dislodge the intruder by making the nest swing violently. Remember that the nest of this species is attached to long and hair-like fungal hyphae. This behaviour can also be elicited experimentally by putting a small ant on the nest while the wasps are away or resting inside their cells. Once they detect the intruder, the wasps immediately leave the nest and fly around it keeping a certain distance. Then one wasp lands at full speed on the nest causing it to swing violently; after one or more attempt the ant eventually falls off. In other cases I noted a wasp first landed on the second cell from the top, and using the first cell as a pivot, moved its abdomen which it held high and slightly bent, back and forth in rapid succession until the nest started to swing in an arc of about 60°. On another occasion I observed a wasp on a nest of only a few cells, spin her nest by wagging her abdomen and by flitting from one part of the nest to another, including the drop stopper. In about 10 min she made the nest swing and spin 28 times, each action lasting 3–4 s. On the 28th attempt the ant fell off. I have seldom seen a wasp attack an ant directly.

While *Metischnogaster* uses its nest as a lever to dislodge ants, various species of *Eustenogaster* seem to use their nests in intra- and interspecific communication. Insects perceive sounds as mechanical vibrations borne on the air, water or substrate. Thanks to its physical properties, the nest carton of social wasps seems to be a good substrate for producing sounds, sometimes audible even to the human ear (Starr 1991). Acoustic signals have been reported in Vespinae and independent-founding Polistinae. Body oscillations producing nest substrate vibration are involved either in adult-larval communication (see Starr 1991; Keeping 1992: Savoyard et al. 1998) or in adult dominance–subordinance interactions (Gadagkar and Joshi 1984; Tindo et al. 1997). The carton vibrations transmit alarm in some species of swarm-founding Polistinae (West-Eberhard 1982) and larvae of species of the genus *Vespa* generate hunger signals by scraping the walls of the cells with their mandibles (Ishay et al. 1974). In hover wasps, sound production is known in *Eustenogaster* which uses the particular structure of the nest envelope as a drum. The phenomenon has been studied by Mike Hansell (1987a) and by my group in *E. calyptodoma* (unpublished data) and in *E. fraterna* (Francescato et al. 2002, unpublished observations) (see Chap. 5, p. 140). The first species is particularly interesting because, as previously noted in the description of the nest, it is the only

Fig. 6.73 *Metischnogaster drewseni* uses the nest as a tool for getting rid of small walking predators

species in which the envelope is completely separated from the cell comb. It has been hypothesised that this architectural solution was a response to selective pressure by parasitoids which drill the cells with their ovopositors to lay eggs on the wasp larvae (Hansell 1987a). However, I believe that the particular configuration of the envelope has also been shaped by the necessity to communicate with conspecifics and congeneric members. A not yet completely verified, but certainly stimulating hypothesis, is that the envelope is a "resonator" which amplifies the beating thus rendering it more efficient in signalling nest occupancy. The intensity of beating could also provide information on the strength of the signalling individual, and the signal could be increased by optimum shape and size of the envelope (see Fig. 5.4). This explains the very high rate of nest usurpation observed throughout the genus and the severe competition for nest use in these wasps, as it represents such an important resource for them also on account of the remarkable energy expenditure necessary for their construction. Unfortunately, as remarked in Chap. 5, this hypothesis remains to be confirmed.

6.6 The Evolution of Nest Architecture

The study of the architecture of animal constructions can be useful for the deduction of phylogenetic relationships (Ducke 1914) and for demonstrating the evolution of adaptive behaviours (Jeanne 1975). In this regard, Wenzel (1991, p. 508) states that "the Stenogastrinae present a promising frontier for future research".

No attempt to reconstruct the phylogenesis of the hover wasps on the basis of nest architectural features exists with the exception of the study by Ohgushi et al. (1990). The authors take into consideration six characters (1) nest material, (2) attachment of nest to substrate, (3) presence of ant guards, (4) spatial orientation of cells, (5) cell arrangement and (6) presence of envelopes. The authors premise, however, that "diversity or the lack of common features (in nest architecture) is a rule in the Stenogastrinae".

The cladogram the Japanese authors obtained using nest architecture characters seems to well fit that proposed by Carpenter (1982), based on morphological and behavioural characters, to the same extent as the one which Wenzel (1993) developed using nest architecture of Polistinae wasps fitted the one proposed for their morphological characters (Carpenter 1991). On the other hand, some critical considerations are necessary, especially with regard to the position of the genus *Liostenogaster*. As already seen, out of all the subfamily, this genus seems to present the widest range of architectonical solutions, some of which also resemble those found in other genera as far as all the characters examined by the Japanese entomologists are concerned. For its peculiar morphological characters this genus is considered at the root of all the hover wasps (Carpenter 1988).

Carpenter (1991) suggests that the common ancestor of the social Vespidae may have already possessed the capability of using vegetable fragments. As Jeanne hypothesises, the simplest architectural scheme for a multi-cellular nest, which does not take into account the saving of construction material, would be a bunch of cylindrical cells with no walls in common and with their bases attached to a flat substrate. A nest described by Ohgushi et al. (1983) with five mud cells and no wall sharing could fit this model but could not be ascribed with certainty to the Stenogastrinae, since no resident wasp was found on it. Recently, however, I found examples of *Liostenogaster* nests in Malaysia which resemble the one described by the Japanese authors. The same can be said for a nest with the cells arranged like organ pipes probably pertaining to an undescribed species of *Liostenogaster* found in central Malaysia (Fig. 6.74). However if these nests can look "primitive" as regards cell assemblage and attachment to the substratum, they are not built with vegetable material, a quality we would expect to find as an ancestral character of the Vespids. If the hypothesis of Hansell that *Liostenogaster*

Fig. 6.74 A nest probably belonging to a species of *Liostenogaster* of the same type as that reported by Ohgushi et al. (1983)

began to use mud again in response to predatory pressure from hornets is correct, this would represent an autapomorphic character of some species of the genus and would be uninformative about their phylogenesis.

If saving material (which leads to the construction of bunches of wall-sharing cells which assume the characteristic hexagonal shape) and the construction of protective structures are to be considered as evolved characters in wasp architecture (as proposed by Jeanne 1975), then the most primitive of all the architectural forms found in the group could be considered to be the nests of a species belonging to the *Parischnogaster jacobsoni* group. As we have seen, these nests consist of sets of cells of vegetable matter scattered over the underside of a leaf, or in a line along a long stem, with no wall-sharing (Fig. 6.39). However, the genus *Parischnogaster* is considered as one of the most derived genera of all hover wasps (Carpenter 1991). The truth is that the architecture of the nest, being directly tied to behaviour, is a character which is highly variable both at intra and inter-specific levels.

It is interesting how various genera have reached different (but functionally equivalent) architectural solutions to common ecological problems—for example, the independent evolution of envelopes and pseudo-envelopes to protect the nest. On the other hand, identical architectural features have also arisen independently in response to the same ecological pressures, such as the viscous ant guards constructed from an abdominal secretion. Responding to the problems Wenzel raised (1991) regarding the great variation in stenogastrine nests with respect to other social vespid taxa, I believe that the lack of a standardised structure of attachment, such as the pedicel, was also a highly important factor which expanded the range of architectural solutions. Poor quality material in part constrained the evolution of large-sized colonies, like other social wasps (but consider the enormous mud nests of *Polybia emaciata*) and this, on its own, contributed to the evolutionary choice of a colony defence strategy mainly based on the small size and camouflage of the nests.

Concerning the development and ontogenetic variation in nest architecture of the various taxa, Wenzel (1991) observes that in Vespinae and Polistinae characters expressed at later stages of nest construction often define lower taxonomic units, while characters expressed at early stages define higher units. The fact that all the stenogastrine wasps initiate nests without a pedicel, for example, differentiates them from all other social wasps. Instead the embellishment of the nest structures should be concentrated in "late" traits and thus are uninformative on phylogenetic distances.

Even during the earliest stage of building, behaviour is not entirely stereotyped; indeed, the changing structure of the nest provides further stimuli to the builders. This system of building organisation based on local stimuli contributed to the evolution of this behaviour in all the animal kingdom (Hansell 2008). The growing structure not only offers new stimuli but also imposes constraints, which direct the assembly process towards a functional outcome.

A good example of development of nest construction that can shed light on the relationships between species of *Liostenogaster* is the nest of *L pardii* which suggests a possible derivation of its architecture from a nest similar to that of *L. flavolineata*. Implantation of the first cells is almost the same in the two species and the origin of the massive shape of *L. pardii* nests evidently derives from the development of the external cells and prolongation of the walls of the nest of *flavolineata* (Fig. 6.17).

In conclusion, a cladistic approach to the development of a phylogenetic tree of the hover wasps based on nest architecture characteristics does not seem practicable at present, even though our knowledge on the nests of the subfamily has much improved since the paper by Ohgushi et al. (1990).

6.6.1 Nests and Speciation

When considering separating species on the basis of nest architecture, we must remember that the nest represents a sort of "frozen behaviour" and its shape and material give important information for the study of the behavioural characters of a certain species. In social wasps in general, and in hover wasps in particular, the nest may be a necessary factor to distinguish different species having a very similar morphology. *Eustenogaster calyptodoma* and *E. micans* and *Parischnogaster alternata* and *P. striatula*, for example, are quite similar from the morphological point of view but quite different in their nest architecture. In our study on *P. alternata*, we also realised how the intra-specific variability of nest architecture could become an important factor in sympatric speciation. In fact we observed how the attachment of the nest to the substrate (flat or thread-like) influenced not only the amount of material necessary for the construction but also the size of the cells. This fact is important as the dimensions of the cells could affect the size of the immature brood and that of the adults. This could eventually bring about possible reproductive isolating mechanisms based on mate choice influenced by the different size of the sexes.

The concept of "ethospecies" has been criticised (cf. Atz 1970, but see Rendall and Di Fiore 2007) and the definition of a species on the basis of its nest alone can be rather feeble; however, Hansell (2008) notes how the building process (and the building behaviour itself) can be considered as equivalent to the molecular and cellular dynamics which characterise the development of an organism. Neither the partial nor the final products of the building process are the behaviour "in itself", but both provide evidence which can substitute the occurrence of the entire event. The result is that we have the great advantage of being able to examine and study the final product (the nest) at leisure and in detail.

References

Atz JW (1970) The application of the idea of homology to behavior. In: Aronson LR, Tobach E, Lehrman DS, Rosenblatt JS (eds) Development and evolution of behavior: essays in memory of TC Schneirla. WH Freeman, San Francisco

Baracchi D, Hashim R, Teseo S, Turillazzi S (2009) Basic social biology and nest architecture of *Liostenogaster topographica* Turillazzi 1999 (Hymenoptera Stenogastrinae). Trop Zool 22:15–25

Barthélémy CA Provisional guide to the social vespids of Hong Kong (Hymenoptera Vespidae). On line: http://insectahk.com/book_page.html

Batra LR, Batra SWT (1966) Fungus-growing termites of tropical India and associated fungi. J Kansas Entomol Soc 39:725–738

Bongiovanni V (1998) L'architettura del nido ed il materiale da costruzione nelle vespe Stenogastrinae (Hymenoptera: Vespidae). Msc Thesis, Università di Firenze, Italy

Carpenter JM (1982) The phylogenetic relationships and natural classification of the Vespoidea (Hymenoptera). Syst Entomol 7:11–38

Carpenter JM (1988) The phylogenetic system of the Stenogastrinae (Hymenoptera: Vespidae). J N Y Entomol Soc 96:140–175

Carpenter JM (1991) Phylogenetic relationships and the origin of social behavior in the Vespidae. In: Ross KG, Matthews RW (eds) The social biology of wasps. Cornell University Press, Ithaca, NY

Coster-Longman (1998) Ecological factors in the agglomeration and evolution of sociality in the Stenogastrinae (Hymenoptera Vespidae). PhD thesis, Università di Firenze, Italy

Coster-Longman C, Turillazzi S (1995) Nest architecture in *Parischnogaster alternata* Sakagami (Vespidae, Stenogastrinae), intra-specific variability in building strategies. Insect Soc 42:1–16

Coster-Longman C, Yong HS, Turillazzi S (1996) Nest architecture in species of the "*Parichnogaster striatula*" group (Hymenoptera, Stenogastrinae). Insect Soc Life 1:77–83

Coster-Longman C, Landi M, Turillazzi S (2002) The role of passive defense (selfish herd and dilution effect) in the gregarious nesting of Liostenogaster wasps (Vespidae Hymenoptera Stenogastrinae). J Insect Behav 15:331–350

Dawkins R (1982) The extended phenotype. Freeman, Oxford

Ducke A (1914) U¨ber Phylogenie und Klassification der sozialen Vespiden. Zoologischen Jahrbu¨chern, Abteilung fu¨r Systematik. Geographie und Biologie der Tiere 36:303–330

Eisner T, Johnessee JS, Carrel J, Hendry LB, Meinwald J (1974) Defensive use by an insect of a plant resin. Science 184:996–999

Emerson AE (1938) Termite nests. A study of the phylogeny of behavior. Ecol Monog 8:247–284
Fanelli D, Boomsma JJ, Turillazzi S (2005) Multiple reproductive strategies in a tropical hover wasp. Behav Ecol Sociobiol 58:190–199
Field J, Foster W, Shreeves G, Sumner S (1998) Ecological constraints on independent nesting in facultatively eusocial hover wasps. Proc R Soc B 265:973–977
Field J, Shreeves G, Sumner S (1999) Group size, queuing and helping decisions in facultatively eusocial hover wasps. Behav Ecol Sociobiol 45:378–385
Francescato E, Massolo A, Landi M, Gerace L, Hashim RB, Turillazzi S (2002) Colony membership, division of labour and genetic relatedness among females of colonies of *Eustenogaster fraterna* (Hymenoptera, Vespidae, Stenogastrinae). J Insect Behav 15:153–170
Gadagkar R (1991) *Belonogaster*, *Mischocyttarus*, *Parapolybia*, and independent founding *Ropalidia*. In: Ross KG, Matthews RW (eds) The social biology of wasps. Cornell University Press, Ithaca
Gadagkar R, Joshi NV (1984) Social organisation in the Indian Wasp, Ropalidia cyathiformis (Hymenoptera, Vespidae). Z Tierpsychol 64:15–32
Hansell MH (1981) Nest construction in the subsocial wasp *Parischnogaster mellyi* (Saussure) (Stenogastrinae Hymenoptera). Insect Soc 28:208–216
Hansell MH (1982) Brood development in the subsocial wasp *Parischnogaster mellyi* (Saussure) (Stenogastrinae Hymenoptera). Insect Soc 29:3–14
Hansell MH (1984) Animal architecture and building behaviour. Longman, London
Hansell MH (1986) Colony biology of the stenogastrine wasp *Holischnogaster gracilipes* (Van der Vecht) on Mount Kinabalu. Entomol Mon Mag 122:31–36
Hansell MH (1987a) Elements of eusociality in colonies of *Eustenogaster calyptodoma* (Sakagami & Yoshikawa) (Stenogastrinae, Vespidae). Anim Behav 35:131–141
Hansell MH (1987b) Nest building as a facilitating and limiting factor in the evolution of eusociality in the Hymenoptera. In: Harvey PH, Partridge L (eds) Oxford surveys in evolutionary biology. Oxford University Press, Oxford
Hansell MH (1996) Wasps make nests: nest make conditions. In: Turillazzi S, West-Eberhard MJ (eds) Natural history and the evolution of paper wasps. Oxford University Press, Oxford
Hansell MH (2008) Built by animals: the natural history of animal architecture. Oxford University Press, Oxford
Hansell MH, Turillazzi S (1990) *Anischnogaster*: from little to less sociality (Stenogastrinae) (Hymenoptera, Vespidae). In: Veeresh GK, Mallik B, Viraktamath CA (eds) Social insects and the environment. Oxford & IBH Publishing Co, New Delhi
Hansell MH, Turillazzi S (1991) A new stenogastrine nest from Papua New Guinea probably belonging to the genus *Stenogaster* Guerin 1831 (Hymenoptera Vespidae). Trop Zool 4:81–87
Hansell MH, Turillazzi S (1995) Nest structure and building material of three species of *Anischnogaster* (Vespidae, Stenogastrinae) from Papua New Guinea. Trop Zool 8:203–219
Hansell MH, Samuel C, Furtado JI (1982) *Liostenogaster flavolineata*: social life in the small colonies of an Asian tropical wasp. In: Breed MD, Michener CD, Evans HE (eds) The biology of social insects. Westview Press, Boulder
Hashim R, Starr CK, Turillazzi S (2006) Nest and species description of the Southeast Asian hover-wasp *Eustenogaster gibbosa* n. sp. (Hymenoptera Vespidae). Trop Zool 19:289–296
Irman AJ, Krebs J (1988) Predation and group living. Trends Ecol Evol 2:31–32
Ishay J, Motro A, Gitter S, Brown MB (1974) Rhythms in acoustical communication by oriental hornet, *Vespa orientalis*. Anim Behav 22:741–744
Iwata K (1967) Report of the fundamental research on the biological control of insect pests in Thailand. II. The report on the bionomics of Aculeate wasps bionomics of subsocial wasps of Stenogastrinae (Hymenoptera, Vespidae). Nat Life SEA 5:259–293
Jeanne RL (1975) The adaptiveness of social wasp nest architecture. Q Rev Biol 50:267–287
Jeanne RL (1979) A latitudinal gradient in rates of ant predation. Ecology 60:1211–1224
Jeanne RL, Bouwma AM (2002) Scaling in nests of a social wasp: a property of the social group reference. Biol Bull 202:289–295

Keegans SJ, Morgan DE, Turillazzi S, Jackson BD, Billen J (1993) The Dufour gland and the secretion placed on eggs of two species of social wasps. J Chem Ecol 19:279–290
Keeping MG (1992) Social organization and division of labour in colonies of the polistine wasp, *Belonogaster petiolata*. Behav Ecol Sociobiol 31:211–224
Krombein KV (1991) Biosystematic studies of Ceylonese wasps. XIX: natural history notes in several families (Hymenoptera, Eumenidae, Vespidae, Pompilidae, Crabronidae). Smithsonian Contrib Zool 515:1–41
Kudô K, Yamane S, Miyano S (1996) Occurrence of a binding matrix in a nest of a primitively eusocial wasp, *Eustenogaster calyptodoma* (Hymenopera, Vespidae). Jpn J Entomol 64:891–895
Landi M, Coster-Longman C, Turillazzi S (2002) Are the selfish herd and the dilution effects important in promoting nest clustering in the hover wasp *Parischnogaster alternata* (Stenogastrinae Vespidae Hymenoptera)? Ethol Ecol Evol 14:297–305
Maschwitz U, Dorow WHO, Botz T (1990) Chemical composition of the nest walls, and nesting behaviour, of *Ropalidia (Icarielia) opifex* van der Vecht, 1962 (Hymenoptera: Vespidae), a Southeast Asian social wasp with translucent nests. J Nat Hist 24:1311–1319
Matsuura M (1991) *Vespa* and *Provespa*. In: Ross KG, Matthews RW (eds) The social biology of wasps. Cornell University Press, Ithaca, NY
O'Donnell S, Jeanne RL (2002) The nest as fortress: defensive behavior of *Polybia emaciata*, a mud-nesting eusocial wasp. J Insect Sci 2:1–5
Ohgushi R, Salmah S (1986) Nests of stenogastrine wasps from two islands of the Mentawai Islands, Indonesia (Hymenoptera, Vespidae). Kontyu 54:561–567
Ohgushi R, Sakagami SF, Yamane S, Abbas ND (1983) Nest architecture and related notes of Stenogastrine wasps in the Province of Sumatera Barat, Indonesia (Hymenoptera, Vespidae). Sci Rep Kanazawa Univ 28:27–58
Ohgushi R, Yamane S, Sakagami SF (1988) Ecological distribution and habitat-linked density of colonies of stenogastrine wasps in tropical S.E. Asia. Zool Sci 5:869–874
Ohgushi R, Sakagami SF, Yamane S (1990) Nest architecture of the stenogastrine wasps: diversity and evolution (Hymenoptera, Vespidae). A comparative review. In: Sakagami SF, Ohgushi R, Roubik DW (eds) Natural history of social wasps and bees in equatorial Sumatra. Hokkaido University Press, Sapporo
Pagden HT (1958) Some Malayan social wasps. Malayan Nat J 12:131–148
Pagden HT (1962) More about *Stenogaster*. Malayan Nat J 16:95–102
Rendall D, Di Fiore A (2007) Homoplasy, homology, and the perceived special status of behavior in evolution. J Hum Evol 52:504–521
Saito F, Kojima JL (2007) A taxonomic revision of the hover wasp genus *Eustenogaster* van der Vecht (Insecta: Hymenoptera; Vespidae, Stenogastrinae). Zootaxa 1556:1–30
Saito F, Nguyen LPT, Carpenter JM, Kojima J (2006) A new *Eustenogaster* species (Hymenoptera: Vespidae; Stenogastrinae), the first hover wasp known to overwinter on the nest. Am Mus Nov 3534:1–12
Sakagami SF, Yoshikawa K (1968) A new ethospecies of *Stenogaster* wasps from Sarawak with a comment on the value of ethological characters in animal taxonomy. Annot Zool Jpn 41:77–84
Samuel C (1987) Factors affecting colony size in the stenogastrine wasp *Liostenogaster flavolineata*. PhD thesis, University of Malaya, Kuala Lumpur
Savoyard JL, Gamboa GJ, Cummings DLD, Foster RL (1998) The communicative meaning of body oscillations in the social wasp, *Polistes fuscatus* (Hymenoptera, Vespidae). Insect Soc 45:215–230
Sledge MF, Fortunato A, Turillazzi S, Francescato E, Hashim R, Moneti G, Jones GR (2000) Use of Dufour's gland secretion in nest defence and brood nutrition by hover wasps (Hymenoptera Stenogastrinae). J Insect Phys 46:753–761
Spradbery JP (1975) The biology of *Stenogaster concinna* Van der Vecht with comments on the phylogeny of Stenogastrinae (Hymenoptera: Vespidae). J Aust Entomol Soc 14:309–318

Spradbery JP (1989) The nesting of *Anischnogaster iridipennis* (Smith) (Hymenoptera: Vespidae) in New Guinea. J Aust Entomol Soc 28:225–228
Starr CK (1991) The nest as the locus of social life. In: Ross KG, Matthews RW (eds) The social biology of wasps. Cornell University Press, Ithaca, NY
Strassmann J, Hughes CR, Turillazzi S, Solis CR, Queller DC (1994) Genetic relatedness and incipient eusociality in stenogastrine wasps. Anim Behav 48:813–821
Tindo M, Francescato E, Dejean A (1997) Abdominal vibrations in a primitively eusocial wasp *Belonogaster juncea juncea* (Vespidae: Polistinae). Sociobiology 29:255–261
Turillazzi S (1986) Colony composition and social behaviour of *Parischnogaster alternata* Sakagami (Hymenoptera Stenogastrinae). Monit Zool Ital 20:333–347
Turillazzi S (1988a) Adults and nest of *Liostenogaster vechti* n.sp. (Hymenoptera Stenogastrinae). Trop Zool 1:193–201
Turillazzi S (1988b) Social biology of *Parischnogaster jacobsoni* (Du Buysson) (Hymenoptera, Stenogastrinae). Insect Soc 35:133–143
Turillazzi S (1990a) Social biology of *Liostenogaster vechti* Turillazzi 1988 (Hymenoptera Stenogastrinae). Trop Zool 3:68–87
Turillazzi S (1990b) Notes on the biology, social behaviour and nest architecture of *Metischnogaster drewseni* (Saussure) (Hymenoptera Stenogastrinae). Boll Zool 57:331–339
Turillazzi S (1991) The Stenogastrinae. In: Ross KG, Matthews RW (eds) The social biology of wasps. Cornell University Press, Ithaca, NY
Turillazzi S (1999) New species of *Liostenogaster* van der Vecht 1969, with keys to adults and nests (Hymenoptera Vespidae Stenogastrinae). Trop Zool 12:335–358
Turillazzi S, Carfi S (1996) Adults and nest of *Liostenogaster pardii* n.sp. (Hymenoptera Stenogastrinae). Trop Zool 9:19–30
Turillazzi S, Gerace L (1992) Size and characteristics of colonies of *Eustenogaster fraterna* (Hymenoptera Vespidae Stenogastrinae). Ethol Ecol Evol (Special Issue) 2:65–67
Turillazzi S, Pardi L (1981) Ant guard in the nest of *Parischnogaster nigricans serrei* (du Buysson). Monit Zool Ital 15:1–7
Turillazzi S, Pardi L (1982) Social behavior of *Parischnogaster nigricans serrei* (Du Buysson) (Hymenoptera Stenogastrinae) in Java. Ann Entomol Soc Am 75:657–664
Turner GF, Pitcher TJ (1986) Attack abatement: a model for group protection by combined avoidance and dilution. Am Nat 128:228–240
van der Vecht J (1972) A review of the new genus *Anischnogaster* in the Papua region (Hymenoptera, Vespidae). Zool Meded Leiden 47:240–256
van der Vecht J (1975) A review of the genus *Stenogaster* Guerin (Hymenoptera: Vespoidea). J Aust Entomol Soc 14:283–308
Vellone A (1986) Il nido di *Parischnogaster nigricans serrei* (Du Buysson) (Hymenoptera, Stenogastrinae). Msc Thesis, Università di Firenze, Italy
Wenzel JW (1991) Evolution of nest architecture. In: Ross KG, Matthews RW (eds) The social biology of wasps. Cornell University Press, Ithaca, NY
Wenzel JW (1993) Application of the biogenetic law to behavioral ontogeny: a test using nest architecture in paper wasps. J Evol Biol 6:229–247
Wenzel JW (1996) Learning, behaviour programs, and higher-level rules in nest construction of *Polistes*. In: Turillazzi S, West-Eberhard MJ (eds) Natural history and the evolution of paper wasps. Oxford University Press, Oxford
West-Eberhard MJ (1978) Polygyny and the evolution of social behavior in wasps. J Kansas Entomol Soc 51:832–856
West-Eberhard MJ (1982) Communication in social wasps: predicted and observed patterns, with a note on the significance of behavioral and ontogenetic flexibility for theories of worker "altruism". In: de Haro A, Espadaler X (eds) La communication chez les sociétés d'insectes. Universidad Autonoma de Barcelona, Bellaterra
Williams FX (1919) Philippine wasp studies, part 2: descriptions of new species and life history studies. Bull Exp Sta Hawaii Sugar Planters' Assoc Entomol 14:19–186

Williams FX (1928) Studies in tropical wasps their hosts and associates (with descriptions of new species). Bull Exp Sta Hawaii Sugar Planters' Assoc Entomol 19:1–179

Yamane S, Okazawa T (1977) Some biological observations on a paper wasp, *Polistes (Megapolistes) tepidus malayanus* Cameron (Hymenoptera, Vespidae) in New Guinea. Kontyû 45:283–289

Yoshikawa K, Ohgushi R, Sakagami SF (1969) Preliminary report on entomology of the Osaka City University 5th Scientific Expedition to Southeast Asia 1966. With description of two new genera of stenogasterine [sic] wasps by J. van der Vecht. Nat Lif SEA 6:153–182

Chapter 7
The Evolution of Social Behaviour in Insects and the Hover Wasps

One will never see, except at the cinema, ants as big as elephants or butterflies as large as eagles! However, some insects have circumvented these limits, evolving special kinds of colonies sometimes known as "superorganisms" which transcend the characteristics of single individuals and have made them a prominent form of life in various terrestrial ecosystems: these are the social insects.

The number of species which can be considered in this group is not great, but in biomass and impact on ecosystems the colonies of termites and ants have been among the dominant elements of most of the land habitats for at least 50 million years, while bees and wasps contribute to the pollination of most of the flowering plants (Wilson 1990; Hölldobler and Wilson 1990; Van Mele and Cuc 2000; Chapman and Bourke 2008)

7.1 Eusociality

Animals form societies of various kinds, from simple groups constituted by individuals which gather together only on special occasions (for example for mating purposes or for feeding or to increase survivorship in unfavourable weather conditions) to spectacular organised societies. In the classical definition (Batra 1966, then developed by Michener 1969), eusocial animals are those the groups of which present three main characteristics (1) their members cooperate in rearing and defending immature brood and (2) adults of different generations remain together at least for a part of their life. But the main feature of their colonies (3) is represented by a *reproductive division of labour* between their members, i.e. some individuals reproduce while others do not and help the first ones to reproduce more. Costa and Fitzgerald (1996) report in a table the changing concept of eusociality (=highest expression of sociality) in the course of the twentieth century (Table 7.1.)

S. Turillazzi, *The Biology of Hover Wasps*, DOI 10.1007/978-3-642-32680-6_7,

Table 7.1 Different concepts of the term eusocial, modified from Costa and Fitzgerald 1996

"Social sensu stricto" (Wheeler 1928): Progeny fed and protected by mother; progeny cooperate with mother in brood rearing
"Social" (Michener 1953): One or both parents survive to cooperate with mature off- spring; division of labour
"Social" (Michener 1958): One or both parents survive to cooperate with mature young: division of labour
"Eusocial" (Batra 1966): Nest-founding parent survives to cooperate with a group of mature offspring; division of labour
"Eusocial" (Michener 1969): Matrifilial family groups with individuals of two generations (mothers and daughters); division of labour with more or less recognisable castes
"Eusocial" (Wilson 1971): Group members use the same composite nest. Group members exhibit: cooperative brood care, overlap of generations and reproductive castes
"Eusocial" (Crespi and Yanega 1995): Group members exhibit alloparental care (with facultative versus obligate castes)
"Eusocial" (Gadagkar 1994): Group members exhibit cooperative breeding (with morphological versus behavioural castes)
"Eusocial" (Sherman et al. 1995): Group members exhibit cooperative breeding; reproductive skew varies from 0 to 1
"Eusocial" (Boomsma 2009): Associations of lifetime monogamous parents and offspring to form obligatory sterile eusocial castes

Various intermediate stages of sociality can be defined according to the absence or presence of one or more of these three conditions. Terms such as *subsocial* and *parasocial* (this one further divided into communal, quasisocial and semisocial) indicate such stages in insects and are resumed in the table reported by Gadagkar (2001) (Table 7.2). Eusociality itself is often divided into primitive and advanced owing to the absence or presence of morphologically differentiated reproductive and non-reproductive castes.

However, while some authors prefer to limit the term "eusocial" to the more specialised forms (Crespi and Yanega 1995), de facto including only some insects, others observe that the presence of altruistic behaviours, the most important evolutionary characteristic, is shared by various organisms (including social insects, cooperatively breeding birds and mammals, sponge shrimps, moulds and some bacteria) (Gadagkar 2001). While others such as Wcislo (1997) and Costa (2006) feel that trying to categorise the various aspects of sociality in animals and other organisms prevents us from understanding the range and complexity of cooperative interactions of myriad forms. In any case, I suggest reading the excellent introduction to the book by Costa (2006) for a deeper insight into the question. According to Gadagkar (2001), the discriminator of eusociality can be considered as the presence of an at least partial, or temporary, reproductive division of labour between the members of a group, leading to differentiation between this status and a more evolved situation in which reproductive and non-reproductive individuals are not only behavioural but also morphologically distinct. These, of course, represent forms of an evolutionary dynamic continuum, which should be considered, together with their variability, not only among different systematic groups but also within different populations of the same taxon.

Table 7.2 Terminology used for various kinds of sociality in insects by several authors (from Gadagkar 2001)

Level	Continued care of young	Cooperative brood care	Reproductive caste differentiation	Overlap of generation	Morphological caste differentiation
SOLITARY	No	No	No	No	No
SUBSOCIAL	Yes	No	No	No	No
PARASOCIAL					
Communal	Yes	No	No	No	No
Quasisocial	Yes	Yes	No	No	No
Semisocial	Yes	Yes	Yes	No	No
EUSOCIAL					
Primitive	Yes	Yes	Yes	Yes	No
Advanced	Yes	Yes	Yes	Yes	Yes

7.2 Social Insects

Eusociality mainly occurs in insects, including not only species belonging to the order Hymenoptera but also Termites (recently definitely recognised as social cockroaches (Inward et al. 2007)). Those orders contain species which are highly eusocial while primitively eusocial groups have also been found in aphids (Homoptera) (Aoki 1977a, b) and thrips (Thysanoptera) (Crespi 1992). Only one species of the largest order of insects, the Coleoptera, has so far been recognised as eusocial: the curculionid *Austroplatypus incompertus* (Kent and Simpson 1992).

Other eusocial animals are represented by certain species of Caribbean shrimps of the genus *Sinalpheus* (Duffy 1996, 1998) which inhabit tropical, usually isolated, sponges and in the mammals, by two species of fossorial rodents, the naked mole rat *Heterocephalus glaber* and the damaraland rat *Mixomis damarensis* (Jarvis 1981; Bennet et al. 1988; Jarvis and Bennet 1993). But social behaviour has also been described and studied in a variety of other organisms starting from procariotes and unicellular eucariotes (for a recent review see West et al. 2007).

Eusocial insects have received increased attention over the last forty years with flourishing of sociobiological research. A milestone book by E.O. Wilson, "The Insect Societies", furnished a clear and exhaustive review of the natural history, biology and social behaviour of the main groups and attracted considerable interest in these insects (Wilson 1971). The enormous amount of new knowledge accumulated in recent years on the characteristics and evolution of social phenomena in the animal kingdom has also permitted highlighting of the differences in the social organisation of insects and vertebrates.

Bees and wasps are especially interesting regarding the study of the evolution of social behaviour as the comparative method can be applied to closely related living species with different degrees of social organisation. The development of the evolution of social characteristics can be inferred by comparing the same behavioural characteristics in species which are phylogenetically close to each other. This method, used first by W.M. Wheeler (1923, 1928) to trace the development of social organisation in insects, followed by various others including Michener (1969) for bees and Evans and West-Eberhard (1970) for wasps, allows us to make reasonable hypotheses on the transitions in the course of evolution, as exhibited in their various patterns of social behaviour.

7.3 Evolution of Social Behaviour

7.3.1 The Biological Altruism Problem

As I stressed above, the passage from temporary to permanent sterility in some individuals of a colony was crucial for the evolution of a more specialised sociality. This is the main question which has puzzled researchers ever since the work of Darwin himself who, in a widely cited passage from the Origin, noted that the sterile castes of social insects posed great problems for his theory. The solution of Darwin, similarly, is cited in a large number of papers (and I can't help reporting it here), but it is often interpreted in slightly different ways: "*This difficulty, though appearing insuperable, is lessened, or, as I believe, disappears, when it is remembered that selection may be applied to the family, as well as to the individual, and may thus gain the desired end (...) a slight modification of structure, or instinct, correlated with the sterile condition of certain members of the community, has been advantageous to the community: consequently the fertile males and females of the same community flourished, and transmitted to their fertile offspring a tendency to produce sterile members having the same modification. And I believe that this process has been repeated, until that prodigious amount of difference between the fertile and sterile females of the same species has been produced, which we see in many social insects*". Darwin (1859, pp. 237–238), tells us two important things: first that evolution can be driven by selection at different levels and, second, that consanguinity, and then genetic similarity, played an important role in the evolution of altruistic behaviour.

7.3.2 Importance of Kinship

Seminal studies by Haldane (1932) and Fisher (1958) showed that genetic similarity between group members was definitely important for the origin and maintenance of altruism because a hypothetical gene for sterility must in any case be passed down to the following generation through related, reproductive individuals, an idea later formalised and elaborated by Hamilton (1964). So, helping a relative to reproduce to a higher extent can guarantee that part of the genes of the sterile helper will be represented in the next generation. However, even completely unrelated individuals can benefit from living in groups (for example in the defence against enemies); some members of the group can renounce direct reproduction in the case of sub-optimal environmental situations, physiological handicap or if forced or constrained by more powerful group mates. In this case they can remain in the group in the hope of obtaining the whole colony (and access to direct reproduction!) if the dominant fails.

Queens and workers, reproductives and helpers, can be recognised exclusively by their behaviour in relatively unspecialized societies but are well distinguished by their morphological and anatomical attributes in more specialised social species. Mechanisms which regulate social organisation and the roles of individuals are based on inter-individual relationships which, in small colonies, lead to the formation of dominance hierarchies with individuals ordered in various ranks with the reproductives at the top. In more evolved societies, the destinies of individuals are decided early on in life, at the larval stage; this is the way the fate of a honeybee larva is determined by food, pollen and nectar or royal jelly, supplied by the workers. In larger and more advanced societies regulatory mechanisms are based on chemical substances, pheromones, which determine and coordinate social life. Social context is, in all cases, decisive in moulding the life of individuals—the larger the society, the smaller, in general, the variations permitted in the social organization of the system.

7.3.3 Hamilton and Inclusive Fitness

W.D. Hamilton defined the importance of relatedness in the evolution of altruism, or self-costing aid; starting in 1964 (the year his two most important papers were published) studies on social insects flourished and began to appear in scientific journals. The concept of inclusive fitness and the theory of kin selection became the model to falsify in an increasing number of experimental investigations. The observation by Fisher and Haldane that an individual can pass on copies of its genes to the following generation not only through direct reproduction but also by helping relatives to increase the number of their brood, became the concept of the term inclusive fitness (Hamilton 1964). In the case of social Hymenoptera then, Hamilton also stressed the fact that the peculiar diploid and haploid egg development and sex determination results in a particularly high degree of relatedness between sisters and a co-respective low relatedness between diploid sisters and haploid brothers. Now haplodiploidy is no longer considered to be fundamental in the social evolution of insects, but this does not lessen the important role of kin selection, as suggested in some recent papers (Novak et al 2010, Hunt 2007; but see also the replies by Abbot et al. 2011; Boomsma et al. 2011; Strassmann et al. 2011; Ferriere and Michod 2011).

Hamilton's famous formula includes both of the two different aspects of social evolution in its two terms: the genetic one (relatedness) and the "ecological" one (costs and benefits of the altruistic act in terms of fitness lost and gained). Biomolecular techniques, which have developed and improved more and more over the last 40 years, have permitted more in depth study of the first term in the equation, i.e. the role of genetic similarity in the evolution of altruism. These studies have demonstrated how kin selection represents a powerful engine in the evolution of sociality and allowed very successful predictions on behaviour in social insects, explaining a wide range of phenomena including altruistic helping, policing, levels of cooperation, sex allocation, etc. (see references in the contributions from Abbot et al. 2011; Boomsma et al 2011; Strassmann et al. 2011; Ferriere and Michod 2011). Owing to difficulties in measuring the actual gain and loss in direct or indirect fitness in various animals, researches which focussed on the second term of the equation were less common but at the same time highly necessary (West-Eberhard 1975; Gadagkar 2001; Costa 2006; Hunt 2007).

7.3.4 Hypotheses on Social Evolution in Insects

In the meantime, several other evolutionary explanations for eusociality in insects have been proposed, including that of Lin and Michener (1972) which laid decisive importance on the associations of non-related individuals in groups determined by ecological needs. Alexander (1974) suggested that sterile individuals can be determined by influential individuals, particularly the parents, by manipulating their physiology; this would thus have favoured reproduction by some other offspring (parental manipulation). Similarly, West-Eberhard (1975, 1978) focussed on the role of mutual relationships, favoured by close proximity, in the evolution of sociality and on the gains in indirect fitness of females with reduced reproductive potential and scarce hopes of obtaining sufficiently high direct fitness (sub-fertility hypothesis). Gadagkar defines this last as a "gambling" stage, an early phase in social evolution in which incentive for group living comes especially from mutualism and reciprocal altruism, made possible by ample complex behavioural repertoires of the individuals involved. In this situation the roles of group members (higher or lower reproductive outputs) are decided by chance (Gadagkar 2001). This stage is followed by a more complex one in which even small differences in the reproductive potentials of the associated females or the capacity of some individuals to influence this potential in others can promote a rise in reproductive skew within the colony (manipulation stage). At the end of the evolutionary process, Gadagkar identifies a substantial increase in the importance of genetic asymmetries which allow preferential altruism towards closer kin. This is, of course, made possible by the efficient mechanisms of recognition (recognition stage) (Gadagkar 2001).

7.3.5 The Evolution of Caste Polymorphism

The passage from primitive to advanced eusociality is not merely a matter of increasing complexity in the differentiation of reproductive and non-reproductive roles in individuals of a social group; it involves decisive changes in gene expression. The genomes of individuals living in a eusocial group are quite similar, as are the genomes of a male and female of the same species. Nevertheless, the differences in morphological, physiological and behavioural phenotypes of the members of a colony in some species are macroscopic. This brings us to another question posed by Darwin, in the same chapter of the Origin, about the evolution of the striking polymorphism of sterile workers in social insects. As Mary Jane West-Eberhard observes, this is a developmental aspect of social evolution where development means the process by which phenotypes are produced and which involves conditional gene expression, which has to be considered an important part of inheritance (West-Eberhard 1996). Considering an ancestral reproductive cycle of solitary species (which West Eberhard defines as the "ovarian groundplan"), the origin of reproductive castes has to be sought in its decoupling into two parts, one that is expressed in workers and the other in queens. According to Gadagkar this happened when some genes, probably doubled and then released from stabilising selection, could have provoked diversifying evolution when some individuals in a species began to rely on the indirect component of inclusive fitness (the workers) and others continued to rely on the direct component (the queens) (Gadagkar 2001). But divergent evolution of the two castes via decoupling of gene expression (as suggested by West-Eberhard 2005) does not require gene duplication, since *different sets* of expressed pre-existing genes come to differentiate the two castes, workers and queens. At a certain point in social evolution, disruptive selection against intermediates in development led to discrete phenotypes in some species, reaching a situation of no return for reproductive and non-reproductive castes differing in behavioural, physiological and morphological traits.

In a recent paper J. Boomsma (2009) suggests that the obligatory sterile eusocial castes only arose via the association of lifetime monogamous parents (also represented by single-mated queens with a full spermatheca) and offspring: only this status would have favoured the above-cited evolutionary succession owing to the high relatedness between offspring in all social insects. Boomsma (who considers as obligate eusociality only those forms where castes are irreversibly determined early in development and workers do not retain the capacity to disperse and form new colonies) states that only when lifetime monogamy has been established even the slightest group benefit will suffice to induce the transition towards eusociality. In this scenario, cooperative breeders such as most primitively eusocial insects and vertebrates and highly eusocial insects would fall under different domains of social evolution (Boomsma 2009).

7.3.6 Towards a Social Concept of the Organism

Eminent researchers of social evolution in insects have long stressed the important role of selection at the colony level in the evolution of higher eusociality (from Wheeler 1923, and Emerson 1939, to the recent book by Hölldobler and Wilson 2009). In general, however, there was a scarce propensity of researchers to investigate the role of any kind of group selection in social evolution, as well as the practical difficulties in experimental manipulation of very large colonies. In any case, the awareness that selection can act at different levels has increased more and more in recent times. In a useful formal model K. Reeve and B. Hölldobler (2007) trace the steps of evolution towards a superorganism considering the effects of individual and colonial selection in situations where the relatedness between group members, the size of the groups and environmental resource availability are variable. The model furnishes testable predictions about the extent to which intergroup conflicts shape the social evolution of a colony. In this model, Kin selection in any event maintains a determinant role, while Wilson and Hölldobler (2005) and Nowak et al. (2010) try to minimise the importance of this kind of selection in social evolution. According to James Costa (2006), Wilson and Hölldobler "underscore the importance of ecological context in social evolution". Costa suggests that even if kin selection did operate in a social system where familial ties are present, focus on the ecology of sociality may be important to recognise that family-dominated models of societies, even if extremely important, are not the only possible models, particularly with regard to non-eusocial groups.

Parallels between colonies of higher social insects and multicellular organisms have been proposed on several occasions ever since the seminal papers by William Morton Wheeler (1923, 1928), including Seeley (1989, 1995), Moritz and Southwick (1992), Tautz (2009), Wilson and Hölldobler (2005), Hölldobler and Wilson (2009). Boomsma himself (2009) recognises that his lifetime monogamy hypothesis can be applied to highly eusocial colonies of social insects as well as to multicellular organisms (see also Maynard Smith and Szathmary 1997). The concept of superorganism developed by Wheeler, defined as the result of a "similarity approach" by Hamilton et al. (2009), underlines the similarities between a multicellular organism and an insect colony. For example, the single workers of a colony are considered as the somatic cells of an organism while reproductive individuals are seen as the gonads; the nest is considered as the skeleton, the soldiers and the alarm system which assure the defence of the colony represent the immune system of the multicellular organism (Table 7.3). Various authors have criticised this concept (cf Hamilton et al. 2009) and observe how this approach "gives very little guidance about what the relevant similarities or how to capture them". The selection approach to a superorganism, instead, emphasises the colony as a unit of selection and has its roots in the kin and group selection theory (Wilson and Hölldobler 2005; Reeve and Hölldobler 2007). The term superorganism itself is criticised; for example Queller and Strassmann (2009) retain that it would be more correct to speak of actual organisms when treating

Table 7.3 Similarities between the components of a highly integrated social insect colony ("superorganism") and a multicellular organism (D. Kronauer personal communication)

Organism	Superorganism
Cells	Colony members
Organs	Castes
Gonads	Reproductive castes
Somatic organs	Worker castes
Immune system	Defensive castes; alarm defence communication; colony recognition
Circulatory system	Food distribution; trophallaxis; allogrooming (chemical cues)
Sensory organs	Combined sensory apparatus of colony members
Nervous system	Communication and interactions among colony members
Skin, skeleton	Nest
Organogenesis: growth and development of the embryo	Sociogenesis: growth and development of the colony

groups of individuals so integrated and where cooperation between the single units is very high and conflicts very low, as we find in the colonies of some highly social insects. Personally I think, like Queller and Strassmann, that if we consider an organism "a complex unit of selection which presents emergent qualities that are not the sum of the single characteristics of its constituting sub-units which lost, at a certain time of their evolutionary history, the capacity to evolve outside the higher unit they formed", we can in any case use the term "superorganism" to define the first level of selection from where we observe a more complex unit: for example, a multicellular organism could correctly be defined as a "superorganism" by a unicellular organism!

Queller and Strassmann (2009) take a further leap forward in considering the importance of social relationships in the evolution of novelties when they extend the concept of organism to any group of cells, individuals and beyond, where internal conflicts are absent or almost absent and high levels of cooperation exist between the lower units, even if these lack contiguity or are composed of different species at different levels of complexity. Communities like those of leaf cutter ants, where insects, fungi and bacteria cooperate in very tight system can be considered organisms: this obviously opens interesting perspectives in the understanding of living systems.

7.4 Social Wasps and Their Importance in the Study of the Evolution of Sociality

Important hypotheses on the evolution of sociality in the superfamily Vespoidea have been put forward by various authors including Richards (1971), Evans and West-Eberhard (1970), West-Eberhard (1978), Pardi (1980), Cowan (1991), Ito (1993), Gadagkar (2001), Hunt (2007). These studies highlighted two main routes

Fig. 7.1 A cluster of mud cells of an unidentified eumenine wasp tended by various females (Malaysia) (picture by H.T. Pagden from the files of J. van der Vecht)

to sociality in these insects: the *matrifilial (subsocial) route* and the *associative (parasocial) route* (see also Sect. 7.1). The authors demonstrated how these evolutionary scenarios could be equally possible because some current species present behavioural characters which could be considered as thresholds crossed during social evolution. In the escalation of parental care we pass from species in which the females lay their eggs on paralysed prey, to species where a female builds a shelter before capturing a prey on which to lay an egg, and then to species in which the time spent by the mother with her own brood increases as the growing larvae are progressively furnished with prey (in some cases partially masticated by the adult). In a series with increasing complexity, parental care extends as far as the adult phase, while more larvae of different ages can be reared at the same time, in cells close to one another. Then the young imagos can find shelter and alimentary reserves in the maternal nest, which will help them reach the reproductive stage more quickly. If we hypothesise that young adults remain on the maternal nest and help their mother or other close relatives to breed more immature brood, then we have all the important characteristics of an evolved social life.

Possible evidences for the other evolutionary scenario (the "associative route") are more common in bees and there are few examples of communal behaviour in vespids. According to Cowan (1991), these kinds of societies could be "preadapted for the evolution of reproductive and worker castes. . .". The same author, however, stresses the fact that "much of the information about these insects consists of the barest anecdotes". An example from my own experience is that of a common Malaysian eumenine species that builds mud cells in the shape of small pots attached one to another (Fig. 7.1). It is not rare to find clusters of hundreds of these cells built by various females which have not necessarily emerged from the same cluster. In 1991, I observed up to 20 different females visiting a cluster of 189 cells, 87 of which still open and with developing larvae. Various females could be found on the cluster of cells at the same time, but each one took care of her own larvae (unpublished observations).

Ito (1993) reports various examples of mutualistic co-operation in wasp species, with mild intranidal dominance relations and low reproductive skew, occurring under particular environmental conditions in which single-female founding is almost impossible. Ito concludes that this "*associative colony foundation could have evolved in the wet tropics through Lin and Michener's (*1972*) mutualistic aggregation hypothesis, rather than through kin selection based 'altruism'*".

Mary Jane West-Eberhard (1978) has observed nest-sharing groups of females belonging to some species of eumenine wasps, including those of the genus *Zethus* (*Zethus miniatus*), *Xenorhynchium*, and *Montezumia*, where females share a common nest composed of cells built by different single individuals, but often reused by others. In *Zethus miniatus* a female without a cell of her own was observed to rear the larva of another individual. This, and the fact that such nests are evidently founded by a single female and show overlap of generations, plus observations of small worker-containing species of *Mischocyttarus* with limited overlap between generations, led this author to propose the so called "Polygynous family hypothesis" which suggests that a "permanently monogynous stage could have been preceded by a "casteless nest sharing" and "rudimentary caste containing, polygynous stage" (see also Sect. 7.3.4).

Carpenter (1991, p. 27) supports the model proposed by West- Eberhard with cladistic analysis but, considering the very low number of Masarinae and Eumeninae reported with strictly casteless nest sharing, retains that "*that stage does not appear to have been distinct from stage III* (the rudimentary caste containing stage) *in the evolution of eusociality in Vespidae*". More recently, the same author (in litteris) points out that "*All one can say about social behavior in eumenines, then and now, is that it is not relevant to the origin of social behavior in the social wasp clade*". However I think that observations like those reported by West-Eberhard are quite important as they regard species of a group considered basal for the evolution of social vespid wasps; obviously they should be corroborated by further studies on these and other species of the subfamily.

In his recent review, Boomsma (see Sect. 7.3.5) concludes that, at present, no true example of parasocial (non-subsocial) evolution has actually been ascertained in Social Insects (Boomsma 2009). He focuses primarily on the conditions for the evolution of *obligatory* workers (determined in the larval stage), and not on the stages of facultative worker production that were important in the *origin* of eusociality (the transition from solitary life to the specialised stage of fixed alloparental care) where the debate necessarily centres on the nature of less specialised groups. If we focus on the eumenines, according to West-Eberhard (in litteris), comparative studies do not support a sub-social route to sociality, as in this group no sub-social groups have ever been reported. Studies already mentioned above by the same author (West-Eberhard 1978, 1987, 1996) support more a transition involving groups of related short-lived females, not necessarily all of the same generation (i.e. not "semisocial"), and in some cases known to be descended from a single nest foundress ("polygynous family hypothesis"). The mating systems (whether monogamous or polygamous) of such casteless group-living and rudimentarily caste-containing wasps are unknown. Subsequent to the origin of workers (helpers with lifetime sterility, whether determined or biased by larval nutrition, as proposed by Hunt (2007), or due to social interactions in the adult stage, both of which can occur simultaneously), there may have been a "window of monogamy" that facilitated the evolution of *obligate* workers, as proposed by Boomsma (2009).

7.5 The Evolution of Social Behaviour in Hover Wasps

7.5.1 Foreword

I have always been, and still am, an experimental ethologist, well aware that long ago Wilson (1975) foresaw the disappearance of ethology and its merger with the sociobiological complex of disciplines. So I am a member of what may be a vanishing or nearly extinct scientific species. I am not a cladist, a molecular biologist, a chemist or a mathematical modeller. My special and peculiar expertise is that I have worked on social wasps for 40 years, over 30 of which on the social behaviour of hover wasps. With these credentials alone I believe I can offer my small contribution to the understanding of the evolution of social life in wasps, based, as it must be, on comparative studies of behaviour in the field and in semi-natural situations in the laboratory.

In the previous chapters I have given an extensive review of the biology and social life of hover wasps. It is now time to focus on the problem of their social evolution and their importance for the study of social behaviour in general. I have already pointed out that various authors have seen these wasps as a key group for understanding the transitions from solitary to social life, as a link in an evolutionary chain or as representing the lineage where the origin of eusociality in wasps took place (Spradbery 1975; van der Vecht 1977; Pardi 1980; West-Eberhard 1978). These discussions have been waiting for more studies to add to the very little information available on the biology and behaviour of these wasps. Now, 20 years after the last review on this argument (Turillazzi 1991) we know many more things, but these are still insufficient for tracing a clear outline of the sociality in the group. Carpenter (1991), in the same chapter where he tests the "polygynous family hypothesis" of West-Eberhard (1978) on the origin of eusociality, observes that "*the ancestral state of the Stenogastrinae is most parsimoniously inferred to be nest sharing (short term or facultative) with temporary reproductive division of labor*". The information available at that time did not differ substantially from what we know today, especially on the most important issue which centres round alloparental care. What Carpenter summarised about the various genera of hover wasps is still unchallenged: "*...in Stenogaster and Eustenogaster, each newly emerged female may initially perform worker tasks but eventually has the*

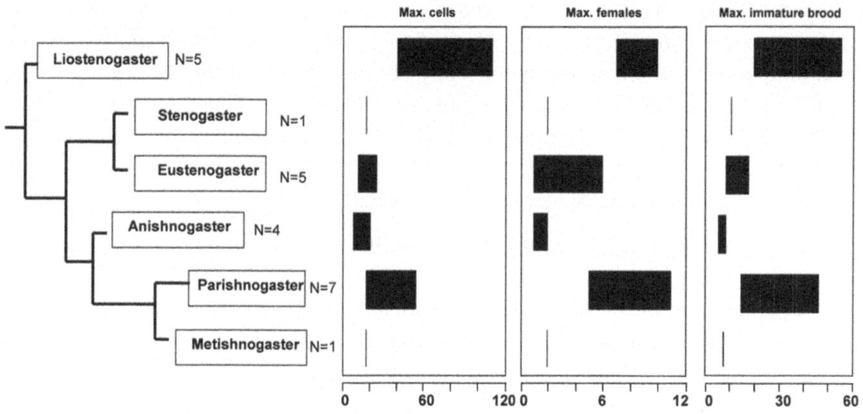

Fig. 7.2 Ranges of maximum number of nest cells, females per colony and immature brood in the nest, for species belonging to six genera of hover wasps (*N* indicates the number of species analysed) (no such data are available for *Cochlischnogaster*). Cladogram of genera is the one proposed by Carpenter (1988)

opportunity to reproduce, either by founding or usurping a new colony or by contending for dominance on the natal nest or in another colony (e.g. Hansell 1987a*). Anischnogaster is apparently similar (Spradbery* 1989*). Liostenogaster has larger colonies and dominance hierarchies (Iwata* 1967*; Hansell et al.* 1982*), as does Parischnogaster (e.g.Turillazzi and Pardi* 1982*). Metischnogaster is poorly known*". More recent studies have generally confirmed this overview, as, at present no one species has been recognised as truly solitary even if a strong tendency for flexible behaviour of single individuals, ranging from solitary to cooperative strategies, has been observed in all the species studied. The description just cited led to the consideration that the hover wasps are examples of the Rudimentary-caste-containing stage, step III in the scenario proposed by M.J. West-Eberhard (1978).

7.5.2 Hover Wasp Sociality

It is not easy to completely characterise the social features of these wasps, given the still imperfect knowledge we have of some genera. In general we can summarise the principal social characteristics of a group by taking into consideration nest size, number of adults (especially females) and immature individuals and, when available, data on sex ratio, presence of reproductive skew, caste dimorphism in the colonies, etc.

Figure 7.2 lists the maximum values for number of cells per nest, adult females, and immature brood per colony reported for a number of species belonging to six genera. The genera are arranged according to the phylogenetic tree proposed by

Carpenter (1988). It is clear that, even if the information on some genera such as *Anischnogaster*, *Stenogaster* and *Metischnogaster* is quite fragmentary, having been obtained from only a few species or just one representative, the highest social "scores" are those for *Liostenogaster* and *Parischnogaster*. These two genera are as distant phylogenetically as they are different in morphology and behavioural repertoires.

The universal features of stenogastrine sociality are as follows:

1. Temporary matrifilial societies and small sized colonies in all the species studied so far.
2. Close interaction between adult and larvae, determined by a particular system of brood rearing based on a special kind of progressive provisioning, which makes the larvae an essential component of the colonies.
3. Presence of, at least temporary and limited, alloparental care confirmed in females of almost all the species studied so far (with the still uncertain situation for *Stenogaster concinna*, Spradbery 1975 and *Anischnogaster laticeps*, Turillazzi and Hansell 1991).
4. Limited reproductive success (defined as the capacity to rear offspring to adult emergence) of isolated nesting females and higher success of nests tended by more than one adult female; this is associated with limited dispersal from the maternal nest and rewards aggregation of individuals (*Liostenogaster, Parischnogaster*).
5. Regulatory mechanisms (dominance hierarchies) which determine high reproductive skew within groups, based on behaviours already present in the repertoires of solitary species and individuals (*Liostenogaster, Parischnogaster, Eustenogaster*)
6. Internidal movement by individual females between maternal and alien nests and alternative behavioural tactics, which may change during the life of a given individual (*Liostenogaster, Parischnogaster, Eustenogaster*).
7. Occurrence of colonies which include unrelated individuals, caused by consistent internidal movements, with variation in intra-colony average relatedness, between adult and immature individuals, ranging from 0.2 to 0.5 (maximum possible offspring–mother relatedness is 0.75 if females mate only once) *Liostenogaster, Parischnogaster, Eustenogaster*).
8. Active mechanisms of nestmate recognition and aggressiveness based on chemical and visual cues which limit entry of migrating alien females into colonies (*Parischnogaster, Liostenogaster*).
9. Colony protection against predators based on passive defence systems, such as camouflage and nest mimicry, or on choice of nesting sites with low predatory pressure.
10. Rudimentary differentiation of reproductive castes, in the females of more socially structured species, determined by physiological mechanisms such as differential ovarian maturation, in time and size (West-Eberhard 1978, 2003; Gadagkar 1991, 1996) (*Liostenogaster, Parischnogaster*).

All these characteristics vary with environmental (including social) conditions, indicating adaptive behavioural flexibility.

7.5.3 *Predisposition to Sociality*

According to Gadagkar (2001) the basic predispositions to eusociality found in the Hymenoptera may be of three types, namely genetic, ecological and physiological-demographic (to which we can add behavioural predispositions, such as contemporary rearing of several larvae in nearby cells, reuse of nests and cells, etc.), and consist in any of the asymmetries that occur in today's primitively eusocial species. Because of their haplodiploid system of sex determination, Hymenoptera can be potentially genetically predisposed to the evolution of eusociality; however, only some of the over 200,000 species of Hymenoptera exhibit more or less advanced social characteristics. Before discussing the various aspects of hover wasp sociality listed above, it is also important to remember that the main factor that facilitated the onset of social groups in some Hymenoptera is the presence of a nest itself, formed by various cells close together, as well as, in wasps, progressive provisioning where larvae of different age are reared at the same time. The presence of a nest and close spatial disposition of the immature brood is a constant feature in social species, including those where the larvae are not walled in their own individual cells, such as ants or allodapine bees. Nests, sometimes complex structures formed by clustered cells, and progressive provisioning are the groundplan for Stenogastrinae, Polistinae and Vespinae, but can also be found in some Eumeninae such as *Zethus*, *Xenorhynchium* and *Montezumia* (recent summary in West-Eberhard 2005). Recently, Hunt stressed the fact that in eusocial Vespids the nest is also a locus rich in proteinaceous food represented by the saliva emitted by the larvae (in Polistinae and Vespinae) and by partly chewed prey brought in by the adults (in Stenogastrinae) (Hunt 2011).

7.5.4 *Social Decisions: Whether or Not to Disperse*

As we have seen above, complex forms of sociality in insects have usually been seen as achieved through two main pathways: the subsocial route and the semisocial route (Wheeler 1928; Evans and West-Eberhard 1970; Michener 1958, 1974; West-Eberhard 1978; Pardi 1980; Wcislo and Tierney 2009). Without the social context created by group formation, individuals of solitary species do not have the opportunity to follow one of the subsequent social trajectories represented by reproductive and helping decisions (Helms Cahan et al. 2002).

Points 1 and 2 of the list in Sect. 7.5.2 are significant for the origin of social behaviour in Stenogastrinae and they also suggest features which may have been important for differentiating the evolution of social life of wasps in general. These follow the first key decision, from an individual perspective, indicated by Helms Cahan et al. (2002), i.e. "whether or not to disperse", which depends on ecological constrains that prevent or limit independent breeding and favour direct or indirect benefits of group living (Stacey and Ligon 1991).

It seems to me totally obvious that in a situation where one female rears a bunch of larvae of slightly different ages all at the same time in cells clustered in a group (which at this point we can call a "colony"), some emerged females may remain with their mother, waiting for the right opportunity to reproduce. Gains in indirect fitness can be only an incidental by-product of group life when the group is composed of relatives (see also Hunt 2011; Boomsma 2009), but undoubtedly they would have favoured the onset of alloparental care. However the same conditions that favour clustering of individual cells in the same place for a single female may also favour clustering of different females in the same place, providing this does not entail over-riding costs (e.g. stealing of provisions and competition for cells) as described in *Zethus miniatus* (West-Eberhard 1978). It is in any case easier to imagine that matrifilial relationships were at the basis of more complex social interactions, because of the presence of an initial, even if temporary, difference in reproductive potential between a mother and her associated offspring, due to the different phases of ovarian development of the mature mother and of the young, still immature, daughter. This would raise the benefit cost ratio for the offspring.

In the case of hover wasps, we can enumerate some of the factors that could have limited the dispersal or promoted the aggregation of individuals. The high failure rate of independent nesting is in general caused by high predation pressure (Turillazzi 1985c; Samuel 1987), but other factors can contribute to making independent nesting more difficult than remaining or joining an existing colony: for example, the scarcity of good places for building a nest (and this is proved by how promptly some species of *Parischnogaster* will use thread substrata experimentally placed in a possible good nesting site [see Sect. 4.2.2)]. Other factors could have attracted individuals to the nest. Reusing cells, by individuals emerged on the same or other nests, would save a lot of energy and avoid the risk of mortality while foraging for nest materials. Reuse of cells is widespread in hover wasps and is especially evident in those species, such as those belonging to the genus *Eustenogaster*, where the nests are particularly elaborate and expensive, in terms of energy, in their construction. One problem regarding cell reutilization could be colonisation of the abandoned cells by various organisms (mainly bark lice and mites, personal observations) which can threaten the health of the wasps' immature brood and the strength of the structure. Once cooperative breeding and alloparental care have evolved, long periods of larval provisioning relative to the length of adult life can also be important in limiting the success rate of a new foundation in species with short-lived adults and nests that take a long time to construct. In this case, the death of one single carer would cause the loss of all the immature, progressively provisioned, brood on the nest. This fact would favour those colonies where more than one adult female cares for the immature brood (but at the same time would favour the evolution of shorter immature developmental periods) as an example of the "assured fitness return" postulated by Strassmann and Queller (1989), Queller (1989) and Gadagkar (1990) and tested in *L. flavolineata* by Field et al. (2000).

Another way of combating predators is colony clustering, as can be observed in the huge aggregations of *L. flavolineata*, *L. vechti* and *P. alternata* nests, but these aggregations are evidently a derived specialisation in the hover wasps.

7.5.4.1 The Larval Component of the Colony and Dufour's Gland Secretion

The nest can also attract adults due to the presence of the immature brood, if this represents a kind of resource. The most striking characteristics of hover wasps are the morphology of the larvae and their peculiar position inside the cells, which allows a unique method of progressive provisioning: the storage of chewed food enclosed within the coiled body of the larva. In at least some polistine and vespine wasps, larval saliva, emitted in large amounts after stimulation of the mouth parts, has been shown to be an attractant for the adults and is composed of a great quantity of free amino acids (Maschwitz 1965; Abe 1991; Hunt 2007) and other substances, some of which, in *Polistes dominula*, with antimicrobial activity (Turillazzi et al. 2006). Larval-adult trophallaxis had already been seen to be an important factor in inducing the origin of social life in these wasps (Roubaud 1916) and this idea, at first doubted by Wilson (1971) and West-Eberhard (1978) due to lack of evidence in the subsocial species observed by Roubaud, has been strongly supported by the findings of Hunt (2007), as an important factor in already eusocial species. The salivary gland in the larvae of hover wasps are not highly developed (see Sect. 2.3.2) and individual larvae artificially stimulated along their pleural lobes extend their bodies, open up their coiled bodies, but do not emit drops of salivary fluid. Moreover, they are perfectly capable of feeding on whole prey owing to their well-developed mandibles (personal observations). However, they can retain boluses of uneaten food as well as drops of liquid in the middle of their coiled bodies and by so doing can accomplish an important function for the colony: acting as food storing devices. The possibility of extra-corporeal digestion by small amounts of saliva cannot be excluded, but careful observations of artificially nourished specimens suggested that this would be trivial, if it occurs at all (unpublished data). Uneaten food is sometimes collected again by adults and used for their own sustenance or to be shared between other larvae (Turillazzi 1985b). It is as yet unclear whether the larvae can prevent the adults from taking food from their coils or not. In any case, I believe that integration of the larval component as an active participant in colony life was an important step in the social evolution of wasps.

The enlarged Dufour's gland found in all stenogastrine wasps is another of their important social characteristics. The production of large amounts of secretion serves various social functions, but it is especially important in rearing the larvae. The matter needs to be discussed in depth as it represents another key innovation in the evolution of these wasps (Turillazzi 1989).

All the species of hover wasps studied up to now present a larval rearing system based on the use of the Dufour's gland secretion. This can be placed on the eggs at the time they are laid, or added after a while and placed on small larvae during their first phases of development. Analysing the data reported in the literature for 16 species belonging to 5 different genera, I found that for a given species the greater the maximum population of adult females or the maximum number of cells in the nest (see values reported in Table 5.1), the lower the proportion of eggs in the nest with no secretion. In other words, in species with larger nests and where there

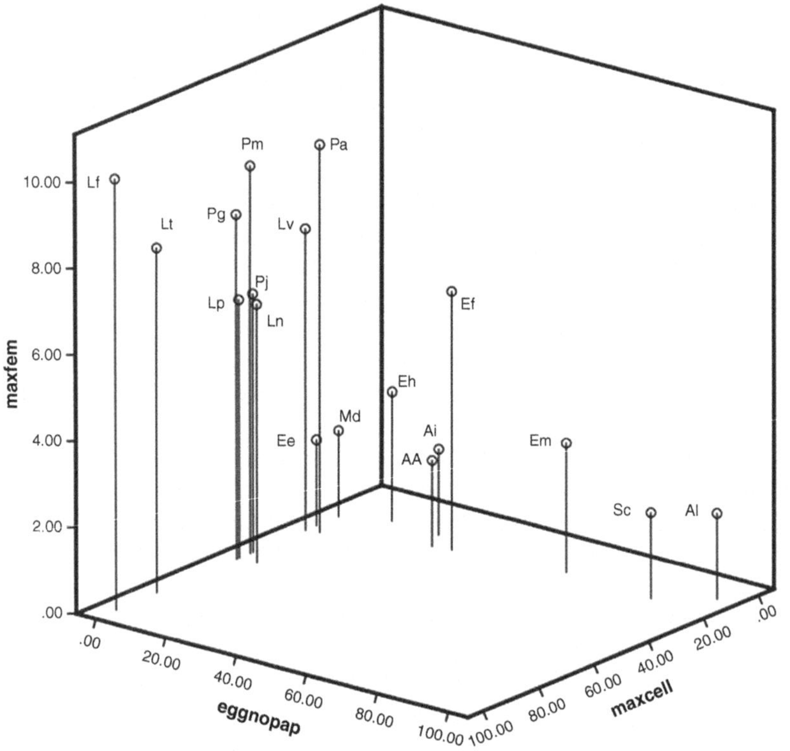

Fig. 7.3 Relation between maximum number of cells and females on the nest and percentage of eggs without Dufour's gland secretion found in nests of 16 species belonging to five genera of Stenogastrinae. Increasing colony size (seen from the *right low corner* to the *upper left one*) corresponds to increasing proportion of eggs supplied with Dufour's gland secretion. AA = *Anischnogaster* sp., Ai = *A. iridipennis*, Al = *A. laticeps*, Em = *Eustenogaster micans*, *Ef* = *E. fraterna*, Ee = *E. eximia*, Eh = *E. hauxwellii*, Lf = *Liostenogaster flavolineata*, Ln = *L. nitidipennis*, Lt = *L. topographica*, Lp = *L .pardii*, Lv = *L. vechti*, Md = *Metischnogaster drewseni*, Pa = *Parischnogaster alternata*, Pg = *P. gracilipes*, Pj = *P. jacobsoni*, Pm = *P. mellyi*, Sc = *Stenogaster concinna*

are generally more females on the nest, the eggs and larvae are generally better supplied with Dufour's gland secretion, the opposite is true for species that characteristically have smaller colonies (percent of eggs without secretion/max no of cells, Spearman ro = −0.727, $P = 0.001$, $N = 18$; percent of eggs without secretion/max no of females, Spearman ro = −0.536, $P = 0.022$, $N = 18$) (Fig. 7.3).

This inverse correlation between colony size (number of adults and cells in the nest) and number of larvae lacking abdominal secretion poses the problem about the basic function of the secretion in the biology of these wasps. As we saw elsewhere (see Sect. 2.2.2.1) this secretion is multifunctional in the various species and is produced by all the females in a colony, whether they are in an egg-laying phase or not, and is one of the most common social activities and aspects of alloparental care. The secretion seems to have no nutritional value, as it is mainly composed of saturated and unsaturated hydrocarbons; although often chewed by adults, they will then discard rather than ingest it. Adults often remove it from cells containing eggs or larvae. Moreover, in some species, such as *P. alternata*, it is quite common to observe recently laid eggs with large amounts of secretion, drops of fluid and pieces of chewed insects that cannot have been deposited for the future larva, because the eggs are far from hatching. All of these observations are consistent with the hypothesis that the secretion represents a substrate or a medium for storing and protecting liquid or malaxated food placed by the adults for the larvae and also as a colonial resource for later distribution.

The invention of this particular kind of progressive provisioning (which could be defined as "storing progressive provisioning") and the extensive, and apparently novel, use of the Dufour's gland secretion could have been important in the evolution of sociality in hover wasps just as larval saliva may have been in the Polistinae + Vespinae: i.e. making the nest an attractive place for adults which find trophic resources on the developing larvae to share with all the individuals of the colony (Turillazzi 1985a, 1989).

Further studies and experimental data are certainly needed to deepen our knowledge of the complex interactions occurring between the adult and the larval components in Stenogastrinae colonies, but it seems likely that these are foundation steps in the evolution of social behaviour in these wasps.

7.5.4.2 From Solitary to Social and Vice Versa

The decision to disperse or remain on the parental nest (or join alien colonies) is an individual choice and alternative behavioural tactics are open at any time to individuals, depending on contingent situations (determined by physiological, environmental and social conditions) that they encounter in the course of their life.

If the alternatives "to disperse–to aggregate" are the first step in the branching social tree of options (Helms Cahan et al. 2002) and flexible behaviour is characteristic of species at the border of sociality, moulded by environmental conditions, then we should expect different reactions, at the population level, under different environmental situations, including variations in the social context (Michener 1969; Wilson 1971). Experiments performed on solitary species, such as the famous one by Sakagami and Maeta (1987) on *Ceratina* solitary bees, demonstrated that in particular situations (high density) females of usually solitary species may show common nest sharing, presence of interactions and reproductive skew. Gadagkar (2001) also suggests that a reverse social evolution could be demonstrated in the case of complete loss of sociality in the descendants of an ancestor that was at least facultatively social. In such a case, morphological traits would be relatively hard to reverse, while behaviour would be relatively easy. Reversions from social to solitary status, inferred from phylogenetic analyses, are not rare in literature; some lineages of halictine bees, for example, show repeated alternation between solitary and eusocial states in different habitats and species of the same genus

(Michener 1964, 1990; Eickwort et al. 1996; Wcislo 1996; Wcislo and Danforth 1997; Danforth 2002). In *Ropalidia formosa* Wenzel (1987) observed a marked fall in group efficiency with respect to other polistine wasps with many females leading an essentially solitary life in dense clusters of separate nests.

Similar experiments and observations have never been made on hover wasps but the phylogenetic position of their genera, proposed by Carpenter (1988), and more recently by Carpenter and Starr (2000), suggests that reversals could occur when considering recent information on their social organisation. Genera such as *Anischnogaster*, and especially *Metischnogaster*, present a far simpler social structure (nests and colonies small in size and precocious departure of newly emerged females) with respect to phylogenetically more basal genera such as *Liostenogaster* and *Parischnogaster* (Fig. 7.2). Maybe these cases represent reversions toward solitary life.

7.5.4.3 Dominance and Subordinance

Social phenomena, such as aggressive interactions, can occur without structured groups in solitary species and are, consequently, easily expressed when animals must interact with other individuals of the same (or different) species. However, advantages tied to dominant positions (especially higher reproductive performance) may not occur immediately and may demand a higher level of social organisation to become appreciable, unless there are obvious differences in physiology and age such as those in "mother–young daughters" pairs. The presence of dominance structured by age is to be expected and, in fact, has been demonstrated in *Liostenogaster flavolineata* (Hansell et al. 1982; Samuel 1987; Bridge and Field 2007); age-related dominance relations are probable in other species of hover wasps that have not yet been extensively studied.

7.5.5 To Breed or Not to Breed

The other basic choice in the decision tree of Helms Cahan et al. (2002) is "to breed or not to breed". This decision is influenced by the ecological, genetic, physiological and social conditions of an individual. In the case of hover wasps it seems that direct reproduction is the default choice for most individuals, which adopt different behaviours, depending on conditions, that ultimately lead to reproduction. The frequent movement of wasps between different colonies and the continuous in-flight nest inspection behaviour observed in various species is clear evidence of the tendency of floating individuals to check opportunities for direct reproduction in the population. Potentially reproductive females evidently wait for an opportunity to reproduce (with eggs just ready to be laid in their ovaries) to exploit any possible weakness of the dominant females. Breeding is the result of a complex game of physiological potentialities and casual exploited opportunities.

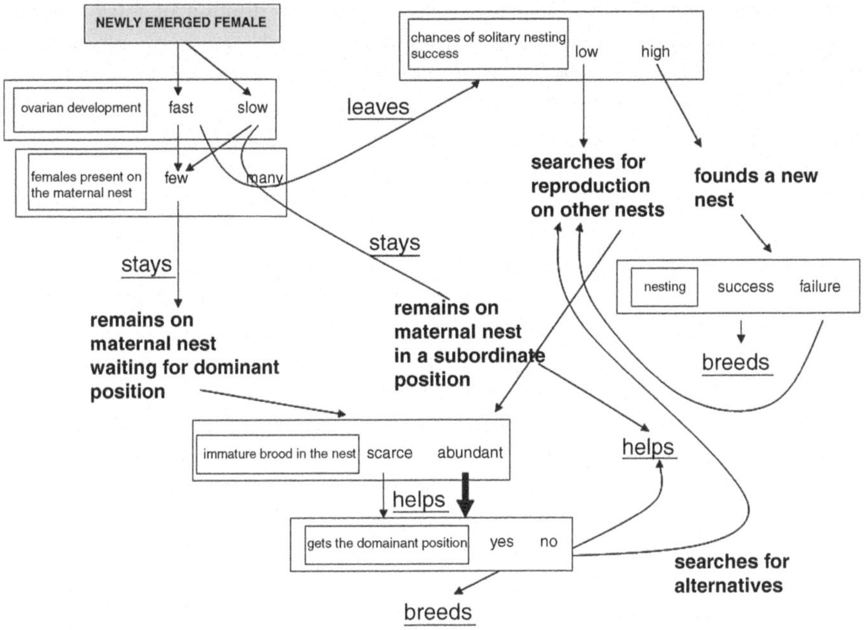

Fig. 7.4 General simplified Decision Tree for a newly emerged female of a species of hover wasp. The flow chart is hypothetical and mainly based on observational data from *Parischnogaster nigricans serrei* (Turillazzi 1985c) and *Liostenogaster flavolineata* (Field 2008). The various decision trajectories must be experimentally checked and are probably different in different species. Boxes represent main factors influencing behavioural decisions

7.5.6 To Help or Not to Help

Helms Cahan et al. (2002) note that organisms can gain benefits from grouping automatically, which include, for example, predator dilution and increased thermoregulatory opportunities. These may require little extra individual investment and may also show low variance in their performance. In contrast, helping or being cooperative towards another member of the group, the fundamental basis of alloparental care, may vary considerably and entail different costs and benefits. In hover wasps, benefits for helpers can be direct or indirect. In the first case if a subordinate female contributes to rearing a large population of adults, this would result in more future helpers to rear the offspring of the subordinate herself, if she becomes dominant and reproductive. Indirect benefits may be gained by the helpers if they rear enough sufficiently related individuals to compensate for the cost to their own reproduction. Helping efforts must be limited and regulated according to the likelihood of the helper surviving to rear her own offspring. All these possibilities have been demonstrated in the hover wasps (Field 2008) and are summarised in Fig. 7.4.

Nestmate recognition mechanisms limit, in part, the dilution of intra-colonial relatedness caused by the internidal drifting of individuals commonly observed in most species. Stenogastrinae seem to use both chemical (based on cuticular hydrocarbons) and visual cues, as observed in the Polistinae and Vespinae. In some cases these cues can also provide information about the fertility of various individuals and could be used to distinguish highly reproductive females or their positions in dominance queues (Zanetti et al. 2001; Cervo et al. 2002; Turillazzi et al. 2004).

In a recent review on the evolution of sociality in wasps, James Hunt affirms that "the origin of worker behaviour, which constitutes the eusociality threshold, is not based on relatedness, therefore the origin of eusociality does not depend on inclusive fitness" (Hunt 2011). On the other hand, Hunt bases his reconstruction of the origin of eusociality on the existence "of single-foundress colonies consisting of a mother and offspring—a family structure...", refusing the semisocial route (Hunt 2011; see also Boomsma 2009). I do not accept the first statement because it contradicts the second, which I am convinced is true (at least for wasps). In fact in the familiar social environment which was presumably where worker behavioural options originated in the first place, this could obviously have been favoured by indirect fitness opportunities. Flexibility over the choice of direct reproduction and alloparental cares towards relatives still exists in living species of hover wasps.

7.5.7 Limiting Factors of the Social Evolution of Hover Wasps

A recurrent question in discussions of social evolution of Stenogastrinae is why these wasps failed to evolve large social colonies of the kind seen in Polistinae and Vespinae. The maximum sizes reached by colonies of these wasps are actually quite small, even with respect to those reached by the smallest colonies of some species of paper wasps. Reasons for this can be various, starting from that proposed by Mike Hansell (1987b) who considered that the quality of the material used by the hover wasps is inadequate for building large nests. Mandibles unsuitable for collecting long fibres of material and poor quality secretion used to assemble the rough pieces of material collected from rotted wood (see also Kudô et al. 1996) may be morpho-functional and architectural constraints. However, as we saw in Chap. 6, the nests of many species are not really fragile, and it is true that even in species where the nest can hold dozens of cells, the adult and larval population is much lower than it could be considering the space available for raising the immature brood. Thus, numerous other factors could be important for limiting the growth in size and complexity of hover wasp societies, including physiological and social ones. First of all, the particular system of oviposition and larval rearing, which implies the production of great quantities of abdominal secretion, could have limited the number of offspring produced by any single female, extending larval development time and raising the energetic costs of parental cares. An adequate production of adults over a short period of time would have furnished a helper force capable of assuring the colony active defence against the main predators; however, immature developmental periods are unusually long compared with those of other social wasps, ranging from 45 to 50 days on average in *Parischnogaster* to more than 100 days in *Liostenogaster*. Active defence, necessary for large colonies for protecting rich sources of food from large predators, is also hampered by the particular use of the sting. This is more suitable for slapping small predators, such as ants, away from the nest and presents morpho-functional constraints for efficient delivery of venom (Fortunato and Turillazzi 2012); furthermore the venom itself does not seem to give a vertebrate much trouble. All these things are probably interconnected and small dimensions and camouflage of the colonies seem to have been the solution pursued by the Stenogastrinae, especially against tropical hornets which prey on nests using visual cues. In the case of *Anischnogaster* and *Stenogaster*, a heavy impact of parasites such as tachinid flies could have caused a heavy reduction in colony size in a sort of reversion to solitary life; while in the case of *Metischnogaster* the elaborate nest design mimicking dry sticks has resulted in an architectonical design which only allows a small number of cells owing to the brittle material.

Jim Hunt (2007, p. 160) recently suggested that "Stenogastrinae never evolved large colony size because there are no diapause females". This evidently did not work for other social vespid taxa which evolved in the same region of the world, such as species of *Ropalidia* and *Polybioides*, but neither in the case of other hover wasps, such as *Eustenogaster nigra* and *E. scitula,* which are dormant when they overwinter (Saito et al. 2009).

7.5.8 Caste Differentiation

Caste differentiation is the process of phenotypic diversification of reproductive and non-reproductive helper individuals. Differentiation may either be imaginal (behavioural and physiological differentiation, occurring in the adult stage) or pre-imaginal (diversification which may include morphological traits, occurring prior to emergence of the adult). Pre-imaginal differentiation is usually considered to be a more specialised, more highly derived state, but may be incidentally involved in differentiation in primitively eusocial species due to factors such as seasonal or individually variable differences in nutrition or growing conditions.

There is no evidence of pre-imaginal differences in the hover wasps; all females are usually totipotent in the sense that they can mate, develop ovaries and lay eggs. Reports of old, unfertilized females in *L. flavolineata*, which could indicate the presence of a worker caste, seem more an exception than the rule (Samuel 1987; Field 2008). However, in both this species and in *P. nigricans serrei* females differ from each other in the timing of fertilisation and onset of ovarian development. This kind of variation could work as a "demographic predisposition" for more complex sociality as it could be advantageous for females with delayed reproductive activity to function as helpers in the nests rather than waiting for full ovarian maturation to found their own colonies (Gadagkar 2001).

7.5.9 Some Considerations on the Evolution of Eusociality in Wasps

I have already stressed that one of the main characteristics of sociality in the Stenogastrinae is the great variability in their social organisation and the great flexibility of individual behaviours on which their variable social organisation is based. They can vary from solitary life to social, even in the same species. This range of variation is not seen in any other subfamily of wasps. This makes the systematic and phylogenetic relationships of the hover wasps of great importance for the study of their biology and the understanding of their evolution.

In a recent paper Pickett and Carpenter (2010) present an analysis of the largest dataset ever used for a phylogenetic treatment of the Vespidae (composed of 267 morphological characters and 66 behavioural characters added to molecular data consisting in 2,943 unaligned sites for three genes) and show convincingly (in opposition to different phylogenetic reconstructions—Schmitz and Moritz 1998 and Hines et al. 2007) that the Stenogastrinae form a clade along with the Polistinae and the Vespinae.[1] From this they conclude that eusociality originated only once in the Vespidae and at the base of that clade. However, as we have already seen in this book the common ancestor of the clade had a worker (helper) phenotype *as one of the options expressed* in a particularly flexible, environmentally sensitive behavioural repertoire, which included options ranging from solitary nesting to participation in common rearing of larvae on maternal or non-parental nests, with possible short periods of latency in direct reproduction. But it is important to note that such a range of options has been found even in a few eumenines such *Zethus miniatus*! (West-Eberhard 1978, 1987). Species presenting this set of possible behaviours are often defined as "facultatively eusocial" (see Hunt 2011 for a recent review). Thus to establish if eusociality evolved once or more times in the Vespidae necessarily depends on a more or less enlarged definition of "eusociality" whereas Pickett and Carpenter (2010) use a narrow definition which requires that eusociality (worker production) is fixed or universal within a given taxon or subfamily. However, this could not have been the case when eusociality originated, necessarily, as an option to solitary life: a novel trait cannot immediately be fixed in the

[1] In particular the results obtained with the dataset which exclude the behavioural characters seem to me decisive (Pickett and Carpenter 2010, Fig 6b, p. 15). First of all because the behavioural characters can be difficult to categorise in two or three alternative states (but see Wenzel 1992 for a compelling argument against this view) and, second, because it could be easy to fall into tautology if we traced the evolution of complex behavioural patterns (but see Deleporte 1993).

lineage of origin, so all novel traits, like the worker phenotype, must begin as an option, or alternative, within a population or species.

The question of the *number* of origins of eusociality in the wasps presents another kind of problems. Although Pickett and Carpenter show convincingly that the Stenogastrinae, Polistinae and Vespinae form a single clade, each subfamily has evolved distinctive forms of sociality with peculiar characteristics, perhaps starting from a different set of relationships between adults and immature brood. For example, colony integration in the Polistinae and Vespinae may have been promoted by larval salivary secretion collected by the adults (Hunt 2007), whereas in the Stenogastrinae it may have been promoted by a particular position and behaviour of the larva and the use of abdominal secretion as a substrate for food reserves. If, on the one hand, we consider these points as the initiators of a more advanced sociality, then the origin of "eusociality" in wasps can be thought of as occurring via two different pathways. On the other hand, if we consider eusociality simply as the presence of a reproductive division of labour between workers and queens, then this "trait" appears to have originated only once. And if we define eusociality on the basis of the presence of irreversible castes (according to Crespi and Yanega 1995) then we have several origins of eusociality in the Vespidae (one or two in the Ropalidiini and one in the Epiponini in the Polistinae and one in the Vespinae). Finally, according to the more recent way of stating the situation proposed by Koos Boomsma (2009), the stenogastrine wasps clade never evolved obligate eusociality, whereas its sister clade, the Polistinae and Vespinae, has only a "single transition" towards obligate eusociality in the ancestor of the Vespines that adopted single queen breeding. It is important to realise that according to this author, cooperative breeding (=facultative eusociality) in hover wasps and also in a number of other wasps and animals, is not separated from solitary breeding by a transition point comparable to the monogamy window, i.e. the point after which obligatory sterile eusocial castes arose via the association of lifetime monogamous parents and offspring (Boomsma 2009). According to the same author, co-operative breeding, including facultative eusociality, and obligate eusociality can be considered different domains of social evolution, because the first are characterised by the continuity of the three variables (genetic and ecological) of Hamilton's equation, while in obligatory eusocial systems relatedness is a class variable. This leads to more complex scenarios of evolutionary ecology in the first systems (Boomsma 2009).

This could well explain not only the similarity but also the differences between the social systems of hover wasps and those of other wasps, insects and even vertebrates. It answers the plea of Wcislo (1997) and Gadagkar (1994, 2001) for a wider consideration and comparative studies of social interactions in several different taxa but, at the same time, it establishes a clear conceptual, even if difficult to be adopted in practice, threshold for the "eusociality", as requested by Crespi and Yanega (1995).

7.6 Future Studies of Hover Wasps

Recently, Jeremy Field (2008) concluded that "the critical ecological factors in the evolution of cooperative breeding and sociality seem to be fundamentally different in wasps and Vertebrates". We have seen, for example, how "insurance advantages" can be quite important in moulding the sociality of some species (*E. fraterna*, *L. flavolineata*) precisely for the particularly extended developmental period of their immatures. This raises the probabilities that a single adult cannot survive long enough to breed after the emergence of its own offspring. This brevity of lifespan is probably rare in social Vertebrates which, instead, seem to be limited in the opportunities for solitary reproduction by the so called "habitat saturation", that is a chronic shortage of breeding territories available for young individuals (Brown 1974, 1978; Emlen 1982; Field 2008). However, given the evident, strong competition for suitable nesting substrata that I observed in some *Parischnogaster* (see Sect. 4.2.2), we can find a similar situation in hover wasps. But the information we have on the demographic aspects of hover wasp biology is still quite scarce and limited to a small set of species; this prevents, at present, a useful comparison between taxa even among the Stenogastrinae themselves.

Future studies on these wasps should give high priority to more extended research on the basic biology and natural history of more species, especially those of the genera in which sociality is limited to a few individuals and to short periods in their life. More attention to the biology, anatomy and physiology of the larvae will cast light on their exact role in the colonial economy.

Owing to their flexible characteristics the colonies of these wasps still remain extremely appropriate subjects to test sociobiological hypotheses, and the high concentrations of some species make it feasible to study sociality at different levels of selection: the individual and the colonial.

Unfortunately the lives of most stenogastrine species are intrinsically connected to the endangered forest environment. This poses serious questions about the future of these delicate and beautiful wasps, as well as for that of myriads of other organisms. Many still unknown species risk disappearing without ever being observed and admired by man. This makes it important for the entomologists living in the countries where these insects are found to assume the task of continued description and study of as many species as they can. I hope that the wonderful example of Charlotte Samuel, a Malaysian citizen who performed the most detailed research ever performed on a hover wasp, will be followed by other young scientists. And I hope that this book will help to stimulate and assist future work on the natural history of the Stenogastrinae in the countries where they live.

References

Abbot P, Abe J, Alcock J, Alizon S, Alpedrinha JAC, Andersson M, Andre JB, Van Baalen M et al (2011) Inclusive fitness theory and eusociality. Nature 471:e1–e4

Abe T (1991) Ecological factors associated with the evolution of worker and soldier castes in termites. Ann Entomol 9:101–107

Alexander RD (1974) The evolution of social behavior. Annu Rev Ecol Syst 5:325–383

Aoki S (1977a) *Colophina clematis* (Homoptera, Pemphigidae), on aphid species with 'soldiers'. Kontyû 45:276–282

Aoki S (1977b) A new species of *Colophina* (Homoptera, Aphidoidea) with soldiers. Kontyû 45:333–337

Batra SWT (1966) Nests and social behavior of halictine bees of India Hymenoptera: Halictidae). Indian J Entomol 28:375–393

Bennet NC, Jarvis JUM, Davies KC (1988) Daily and seasonal temperatures in the burrows of African rodent moles. S Afr J Zool 23:189–195

Boomsma JJ (2009) Lifetime monogamy and the evolution of eusociality. Phil Trans R Soc Lond B 364:3191–3207

Boomsma JJ, Beekman M, Cornwallis CK, Griffin AS, Holman L, Hughes WOH, Keller OBP, Ratnieks FLW (2011) Only full-sibling families evolved eusociality. Nature 471:e4–e5

Bridge C, Field J (2007) Queuing for dominance: gerontocracy and queue-jumping in the hover wasp *Liostenogaster flavolineata*. Behav Ecol Sociobiol 61:1253–1259

Brown JL (1974) Alternate routes to sociality in jays with a theory for the evolution of altruism and communal breeding. Am Zool 14:63–80

Brown JL (1978) Avian communal breeding systems. Annu Rev Ecol Syst 9:123–155

Carpenter JM (1988) The phylogenetic system of the Stenogastrinae (Hymenoptera: Vespidae). J N Y Entomol Soc 96:140–175

Carpenter JM (1991) Phylogenetic relationships and the origin of social behavior in the Vespidae. In: Ross KG, Matthews RW (eds) The social biology of wasps. Cornell University Press, Ithaca, NY

Carpenter JM, Starr CK (2000) A new genus of hover wasps from Southeast Asia (Hymenoptera: Vespidae; Stenogastrinae). Am Mus Novit 3291:1–12

Cervo R, Dani FR, Zanetti P, Massolo A, Turillazzi S (2002) Chemical nestmate recognition in a stenogastrine wasp, *Liostenogaster flavolineata* (Hymenoptera Vespidae). Ethol Ecol Evol 14:351–363

Chapman RE, Bourke AFG (2008) The influence of sociality on the conservation biology of social insects. Ecol Lett 4:650–662

Costa JT (2006) The other insect societies. Belknap Press of Harvard University Press, Cambridge, MA

Costa JT, Fitzgerald TD (1996) Developments in social terminology: semantic battles in a conceptual war. Trends Ecol Evol 11:285–289

Cowan DP (1991) The solitary and presocial Vespidae. In: Ross KG, Matthews RW (eds) The social biology of wasps. Cornell University Press, Ithaca, NY

Crespi BJ (1992) Eusociality in Australian gall thrips. Nature 359:724–726

Crespi BJ, Yanega D (1995) The definition of eusociality. Behav Ecol 6:109–115

Danforth BN (2002) Evolution of sociality in a primitively eusocial lineage of bees. Proc Natl Acad Sci USA 99:286–290, http://www.pnas.org/search?author1=Bryan+N.+Danforth&sortspec=date&submit=Submit

Darwin C (1859) The origin of species by means of natural selection, 1st edn. John Murray, London, http://www.talkorigins.org/faqs/origin.html

Deleporte P (1993) Characters, attributes and tests of evolutionary scenarios. Cladistics 9:427–432

Duffy JE (1996) Eusociality in a coral-reef shrimp. Nature 381:512–514

Duffy JE (1998) On the frequency of eusociality in snapping shrimps with description of a new eusocial species. Bull Mar Sci 63:387–400

Eickwort GC, Eickwort JM, Gordon J, Eickwort MA (1996) Solitary behavior in a high altitude population of the social sweat bee, *Halictus rubicundus* (Hymenoptera: Halictidae). Behav Evol Sociobiol 38:227–233

Emerson AE (1939) Social coordination and superorganism. Am Mid Nat 21:182–209

Emlen ST (1982) The evolution of helping. An ecological constraints model. Am Nat 119:29–39

Evans HE, West-Eberhard MJ (1970) The wasps. University Michigan Press, Ann Arbor
Ferriere R, Michod RE (2011) Inclusive fitness in evolution. Nature 471:e6–e8
Field J (2008) The ecology and evolution of helping in hover wasps (Hymenoptera: Stenogastrinae). In: Korb J, Heinze J (eds) Ecology of social evolution. Springer, Heidelberg, Berlin
Field JP, Shreeves G, Sumner S, Casiraghi M (2000) Insurance-based advantage to helpers in a tropical hover wasp. Nature 404:869–871
Fisher RA (1958) The genetical theory of natural selection, 2nd edn. Dover, New York
Fortunato A, Turillazzi S (2012) Dufour's gland and massive use of its secretion drove the evolution of sting's morphology and function in hover wasps (Hymenoptera Stenogastrinae). Arthropod Struct Dev 41:259–264
Gadagkar R (1990) Evolution of eusociality: the advantage of assured fitness returns. Phil Trans R Soc Lond B 329:17–25
Gadagkar R (1991) *Belonogaster, Mischocyttarus, Parapolybia,* and independent founding *Ropalidia*. In: Ross KG, Matthews RW (eds) The social biology of wasps. Cornell University Press, Ithaca
Gadagkar R (1994) Why the definition of eusociality is not helpful to understand its evolution and what should we do about it. Oikos 70:485–487
Gadagkar R (1996) The evolution of eusociality, including a review of the social status of *Ropalidia marginata*. In: Turillazzi S, West-Eberhard MJ (eds) Natural history and the evolution of paper wasps. Oxford University Press, Oxford
Gadagkar R (2001) The social biology of *Ropalidia marginata*: toward understanding the evolution of eusociality. Harvard University Press, Cambridge, MA
Haldane JBS (1932) The causes of evolution. Longmans, Green and Co, London
Hamilton WD (1964) The genetical evolution of social behavior. J Theor Biol 7:1–52
Hamilton A, Smith NR, Haber MH (2009) Social insects and the individuality thesis: cohesion and the colony as a selectable individual. In: Gadau J, Fewell J (eds) Organization of Insect Societies. Harvard University Press, Cambridge, MA
Hansell MH (1987a) Elements of eusociality in colonies of *Eustenogaster calyptodoma* (Sakagami & Yoshikawa) (Stenogastrinae, Vespidae). Anim Behav 35:131–141
Hansell MH (1987b) Nest building as a facilitating and limiting factor in the evolution of eusociality in the Hymenoptera. In: Harvey PH, Partridge L (eds) Oxford surveys in evolutionary biology. Oxford University Press, Oxford
Hansell MH, Samuel C, Furtado JI (1982) *Liostenogaster flavolineata*: social life in the small colonies of an Asian tropical wasp. In: Breed MD, Michener CD, Evans HE (eds) The biology of social insects. Westview Press, Boulder
Helms Cahan S, Blumstein DT, Sundström L, Liebig J, Griffin A (2002) Social trajectories and the evolution of social behavior. Oikos 96:206–216
Hines HM, Hunt JH, O'Connor TK, Gillespie JJ, Cameron SA (2007) Multigene phylogeny reveals eusociality evolved twice in vespid wasps. Proc Natl Acad Sci USA 104:3295–3299
Hölldobler B, Wilson EO (1990) The ants. Springer, Berlin
Hölldobler B, Wilson EO (2009) The superorganism: the beauty, elegance and strangeness of insect societies. W. W. Norton & Company, New York
Hunt JH (2007) The evolution of social wasps. Oxford University Press, New York
Hunt JH (2011) A conceptual model for the origin of worker behaviour and adaptation of eusociality. J Evol Biol 25:1–19
Inward D, Beccaloni G, Eggleton P (2007) Death of an order: a comprehensive molecular phylogenetic study confirms that termites are eusocial cockroaches. Biol Lett 3:331–335
Ito Y (1993) Behaviour and social evolution of wasps: the communal aggregation hypothesis, Oxford series in ecology and evolution. Oxford University Press, Oxford
Iwata K (1967) Report of the fundamental research on the biological control of insect pests in Thailand. II. The report on the bionomics of Aculeate wasps bionomics of subsocial wasps of Stenogastrinae (Hymenoptera, Vespidae). Nat Life SEA 5:259–293

Jarvis JU (1981) Eusociality in a mammal: cooperative breeding in naked mole-rat colonies. Science 212:571–573

Jarvis JUM, Bennet NC (1993) Eusociality has evolved independently in two genera of bathyergid mole-rats but occurs in no other subterranean mammal. Behav Ecol Sociobiol 33:353–360

Kent DS, Simpson JA (1992) Eusociality in the beetle *Austroplatypus incompertus* (Coleoptera: Platypodidae). Naturwissenshaften 79:86–87

Kudô K, Yamane S, Miyano S (1996) Occurrence of a binding matrix in a nest of a primitively eusocial wasp, *Eustenogaster calyptodoma* (Hymenopera, Vespidae). Jpn J Entomol 64:891–895

Lin N, Michener CD (1972) Evolution of sociality in insects. Q Rev Biol 47:131–159

Maschwitz U (1965) Larven als Nahrungsspeicher im Wespenvolk (Ein Beitrag zum Throphallaxie-problem). Verh Dt Zool Ges 50:530–534

Maynard Smith J, Szathmary E (1997) The major transitions in evolution. Oxford University Press, Oxford

Michener CD (1953) Problems in the development of social behavior and communication among insects. Trans Kansas Acad Sci 56:1–15

Michener CD (1958) The evolution of social behaviour in bees. Proceedings of the Tenth International Congress of Entomology, vol 2, pp 441–447

Michener CD (1964) Reproductive efficiency in relation to colony size in hymenopterous societies. Insect Soc 11:317–341

Michener CD (1969) Comparative social behavior of bees. Annu Rev Entomol 14:299–342

Michener CD (1974) The social behavior of the bees. Harvard University Press, Cambridge, MA

Michener CD (1990) An evolutionary approach to castes and reproduction. Springer, New York

Moritz RFA, Southwick EE (1992) Bees as superorganisms. an evolutionary reality. Springer, Heidelberg

Nowak MA, Tarnita CE, Wilson EO (2010) The evolution of eusociality. Nature 466:1057–1062

Pardi L (1980) Le vespe sociali: biologia ed evoluzione del comportamento. Accad Naz Lincei Contrib Centro Linceo Interdisc Sc Mat Appl 51:161–221

Pickett KM, Carpenter JM (2010) Simultaneous analysis and the origin of Eusociality in the Vespidae (Insecta: Hymenoptera). Arthropod Syst Phyl 68:3–33

Queller DC (1989) The evolution of eusociality: reproductive head starts of workers. Proc Natl Acad Sci USA 86:3224–3226

Queller DC, Strassmann JE (2009) Beyond society: the evolution of organismality. Phil Trans R Soc B 364:3143–3155

Reeve HK, Hölldobler B (2007) The emergence of a superorganism through intergroup competition. Proc Natl Acad Sci USA 104:9736–9740

Richards OW (1971) The biology of the social wasps (Hymenoptera, Vespidae). Biol Rev 46:483–528

Roubaud E (1916) Recherches biologiques sur les guêpes solitaires et sociales d'Afrique. La genèse de la vie sociale et l'évolution de l'instinct maternel chez les vespides. Ann Sci Nat Zool 1:1–160

Saito F, Nguyen LPT, Kojima J (2009) Colony cycle of a "temperate" hover wasp, *Eustenogaster nigra*, with special reference to overwintering of males in an enveloped nest together with virgin females (Hymenoptera, Vespidae, Stenogastrinae). Insect Soc 56:49–54

Sakagami SF, Maeta Y (1987) Sociality, induced and/or natural, in the basically solitary small carpenter bees (*Ceratina*). In: Ito Y, Brown J, Kikkawa J (eds) Animal societies: Theories and facts. Japan Scientific Societies Press, Tokyo

Samuel C (1987) Factors affecting colony size in the stenogastrine wasp *Liostenogaster flavolineata*. PhD thesis, University of Malaya, Kuala Lumpur

Schmitz J, Moritz RFA (1998) Molecular phylogeny of Vespidae (Hymenoptera) and the evolution of sociality in wasps. Mol Phylogenet Evol 9:183–191

Seeley TD (1989) The honey bee colony as a superorganism. Am Sci 77:546–553

Seeley TD (1995) The wisdom of the hive. Harvard University Press, Cambridge, MA

Sherman PW, Lacey EA, Reeve HK, Keller L (1995) The eusociality continuum. Behav Ecol 6:102–108

Spradbery JP (1975) The biology of *Stenogaster concinna* Van der Vecht with comments on the phylogeny of Stenogastrinae (Hymenoptera: Vespidae). J Aust Entomol Soc 14:309–318

Spradbery JP (1989) The nesting of *Anischnogaster iridipennis* (Smith) (Hymenoptera: Vespidae) in New Guinea. J Aust Entomol Soc 28:225–228

Stacey PB, Ligon JD (1991) The benefits of philopatry hypothesis for the evolution of cooperative breeding: habitat variance and group size effects. Am Nat 137:831–846

Strassmann JE, Queller DC (1989) Ecological determinants of social evolution. In: Breed MD, Page RE (eds) The genetics of social evolution. Westview, Boulder

Strassmann JE, Page RE, Robinson GE, Seeley TD (2011) Kin selection and eusociality. Nature 471:e5–e6

Tautz J (2009) The buzz about bees—biology of a superorganism. Springer, Berlin

Turillazzi S (1985a) Function and characteristics of the abdominal substance secreted by wasps of the genus *Parischnogaster*. Monit Zool Ital 19:91–99

Turillazzi S (1985b) Brood rearing behaviour and larval development in *Parischnogaster nigricans serrei*. Insect Soc 32:117–127

Turillazzi S (1985c) Colonial cycle of *Parischnogaster nigricans serrei* (Du Buysson) in West Java (Hym. Stenogastrinae). Insect Soc 32:43–60

Turillazzi S (1989) The origin and evolution of social life in the Stenogastrinae (Hymenoptera Vespidae). J Insect Behav 2:649–661

Turillazzi S (1991) The Stenogastrinae. In: Ross KG, Matthews RW (eds) The social biology of wasps. Cornell University Press, Ithaca, NY

Turillazzi S, Hansell MH (1991) Biology and social behaviour of three species of *Anischnogaster* (Vespidae, Stenogastrinae) in Papua New Guinea. Insect Soc 38:423–437

Turillazzi S, Pardi L (1982) Social behavior of *Parischnogaster nigricans serrei* (Du Buysson) (Hymenoptera Stenogastrinae) in Java. Ann Entomol Soc Am 75:657–664

Turillazzi S, Sledge MF, Dapporto L, Landi M, Fanelli D, Fondelli L, Zanetti P, Dani FR (2004) Epicuticular lipids and fertility in primitively social wasps (Hymenoptera Stenogastrinae). Physiol Entomol 29:1–8

Turillazzi S, Mastrobuoni G, Dani FR, Moneti G, Pieraccini G, La Marca G, Bartolucci G, Perito B, Lambardi D, Cavallini V, Dapporto L (2006) Dominulin A and B: two new antibacterial Peptides identified on the cuticle and in the venom of the social paper wasp *Polistes dominulus* using MALDI-TOF, MALDI-TOF/ TOF, and ESI-Ion Trap. J Am Soc Mass Spectrom 17:376–383

van der Vecht J (1977) Studies of oriental Stenogastrinae (Hymenoptera: Vespoidea). Tijdschr Entomol 120:55–75

Van Mele P, Cuc NTT (2000) Evolution and status of *Oecophylla smaragdina* (Fabricius) as a pest control agent in citrus in the Mekong Delta, Vietnam. Int J Pest Manage 46:295–301

Wcislo WT (1996) Commentary: solitary behavior in social bees. Behav Ecol Sociobiol 38:235–236

Wcislo WT (1997) What are words worth? Trends Ecol Evol 12:161

Wcislo WT, Danforth BN (1997) Secondarily solitary: the evolutionary loss of social behavior. Trends Ecol Evol 12:468–474

Wcislo WT, Tierney SM (2009) Evolution of communal behavior in bees and wasps: an alternative to eusociality. In: Gadau J, Fewell J (eds) Organization of insect societies: from genomes to socio-complexity. Harvard University Press, Cambridge, MA

Wenzel JW (1987) *Ropalidia formosa*, a nearly solitary paper wasp from Madagascar (Hymenoptera: Vespidae). J Kansas Entomol Soc 60:549–556

Wenzel JW (1992) Behavioral homology and phylogeny. Annu Rev Ecol Syst 23:261–281

West SA, Diggle SP, Buckling A, Gardner A, Griffin AS (2007) The social lives of microbes. Annu Rev Ecol Evol Syst 38:53–77

West-Eberhard MJ (1975) The evolution of social behavior by kin selection. Q Rev Biol 50:1–33

West-Eberhard MJ (1978) Polygyny and the evolution of social behavior in wasps. J Kansas Entomol Soc 51:832–856

West-Eberhard MJ (1987) Flexible strategy and social evolution. In: Ito Y, Brown JL, Kikkawa J (eds) Animal societies: theories and facts. Japan Scientific Societies Press, Tokyo

West-Eberhard MJ (1996) Wasp societies as microcosms for the study of development and evolution. In: Turillazzi S, West-Eberhard MJ (eds) Natural history and the evolution of paper wasps. Oxford University Press, Oxford

West-Eberhard MJ (2003) Developmental plasticity and evolution. Oxford University Press, New York

West-Eberhard MJ (2005) Phenotypic accommodation: adaptive innovation due to developmental plasticity. J Exp Zool Part B 304:610–618

Wheeler WM (1923) Social life among the insects. Hartcourt Brace, New York

Wheeler WM (1928) The social insects: their origin and evolution. Kegan, Paul, Trench, Trubner and Co, London

Wilson EO (1971) The insect societies. Belknap Press of Harvard University Press, Cambridge, MA

Wilson EO (1975) Sociobiology: the new synthesis. Belknap Press of Harvard University Press, Cambridge, MA

Wilson EO (1990) Success and dominance in ecosystems: the case of the social insects. Oldendorf/Luhe, Germany

Wilson EO, Hölldobler B (2005) Eusociality: origin and consequences. Proc Natl Acad Sci USA 102:13367–13371

Zanetti P, Dani FR, Destri S, Fanelli D, Massolo A, Moneti G, Pieraccini G, Turillazzi S (2001) Nestmate recognition in *Parischnogaster striatula* (Hymenoptera Stenogastrinae), visual and olfactory recognition cues. J Insect Phys 47:1013–1020

Index

S. Turillazzi, *The Biology of Hover Wasps*, DOI 10.1007/978-3-642-32680-6,

The manufacturer's authorised representative in the EU is Springer Nature Customer Service Centre GmbH, Europaplatz 3, 69115 Heidelberg, Germany. If you have any concerns regarding our products, please contact ProductSafety@springernature.com

Printed and bound by CPI Group (UK) Ltd, Croydon, CR0 4YY
30/09/2026
02225591-0009